OUR ANCIENT LAKES

OUR ANCIENT LAKES

A Natural History

JEFFREY MCKINNON

The MIT Press
Cambridge, Massachusetts
London, England

Grant funding provided by Furthermore: a program of the J. M. Kaplan Fund.

The MIT Press would like to thank the anonymous peer reviewers who provided comments on drafts of this book. The generous work of academic experts is essential for establishing the authority and quality of our publications. We acknowledge with gratitude the contributions of these otherwise uncredited readers.

This book was set in Adobe Garamond Pro by New Best-set Typesetters Ltd. Printed and bound in the United States of America.

Library of Congress Cataloging-in-Publication Data is available.

ISBN: 978-0-262-04785-2

10 9 8 7 6 5 4 3 2 1

For Rene, Tara, and Andrew

Contents

Preface

I think that by retaining one's childhood love of such things as trees, fishes, butterflies and . . . toads, one makes a peaceful and decent future a little more probable.
—George Orwell, 1946

It was in Irkutsk, just down the Angara River from Siberia's Lake Baikal, that I got religion on ancient lakes. The year was 2002, and I had already read many scientific papers about the exotic faunas of Baikal, Tanganyika, Titicaca, and others, and had been conducting my own work on Sulawesi's Lake Matano. But seeing Baikal and attending my first Speciation in Ancient Lakes conference, hosted in Irkutsk, switched a light on and made material the striking differences between ancient lakes and other freshwater systems (figure P.1 shows ancient lakes based on a relatively inclusive definition; for more on definitions of *ancient*, see chapter 1). Baikal, whose age is measured in the tens of millions of years, was sublime, with multicolored shrimplike amphipods teeming in the shallows and candelabra-shaped sponges abundant just a little deeper. The lake seemed endless as we explored a small corner of it on a cruise after the

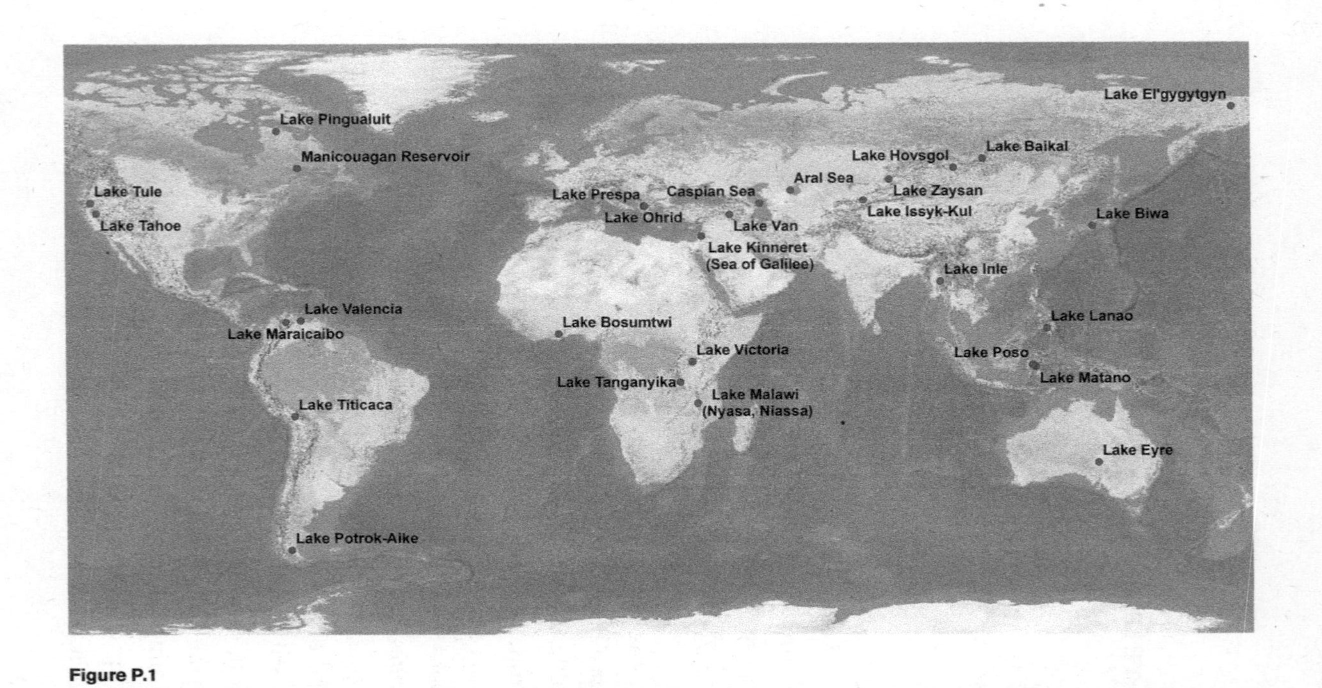

Figure P.1

Locations of ancient lakes as characterized by Stephanie Hampton and colleagues. *Source:* Slightly modified from Hampton et al., "Recent Ecological Change in Ancient Lakes," *Limnology and Oceanography* 63 (2018). Published under a Creative Commons CC BY-NC 3.0 license (https://creativecommons.org/licenses/by-nc/3.0).

Figure P.2
Attendees at the 2022 Species in Ancient Lakes conference in Kigoma, Tanzania, on the shores of Lake Tanganyika. *Source:* Photo by Mupape Mukuli.

conference, in a vessel I would have guessed came from the North Pacific had I seen it in a photo. At the conference, I had heard mesmerizing reports from a community of researchers who were looking at the wildly diverse denizens of these, our planet's oldest bodies of freshwater.

I have since attended several ancient lake conferences (the participants at the last conference before I completed this book are shown in figure P.2) and have always come away inspired. Yet when I get home and talk about the wonders of the lakes with nonconference colleagues, including experts in ecology and evolution who work in marine or terrestrial systems, I am often met with blank stares. Frequently my peers are familiar with the cichlid fishes of Africa's Lake Victoria and their extinction crisis, or have some vague notion that Lake Baikal is old and contains an awful lot of fresh water. But they seldom know what is meant by the term *ancient lake*, or why such lakes matter.

As time passed, my professional frustrations evolved into a deeper concern for the lakes and their often-perilous circumstances. These frustrations and worries came to a head during fieldwork in Sulawesi in the 2000s as I saw development exploding around Lake Matano. I can still hear my old friend Peter Hehanussa, a senior Indonesian scientist, as he pleaded with the management of the nickel mine that dominated the area not to build a road around the lake, so sure was he that it would accelerate the degradation of Matano and its fauna. Hectare after hectare of forest was being cleared, and houses were popping up like mushrooms after a spring rain. I was also hearing of ever more nonnative fishes becoming established—new ones most every year—and knew that the situation was the same or worse in other lakes. It seemed time to attempt something beyond basic research and an occasional outreach effort. I thought that a book might draw more attention to the lakes and communicate their importance, especially if it was not too full of jargon or despair. I resolved to get started on a manuscript.

So how is it that the publication date on this volume is 2023? A job change, move, and new administrative responsibilities came up before I was able to make meaningful progress on my book plan. The manuscript ended up on hold for over a decade until I stepped away from administrative duties. When I finally returned to the project, I quickly found myself of two minds. I was pleased to find that the book I had imagined seemed yet to be needed. But I was also a little sad that ancient lakes were still not broadly appreciated, despite the efforts of my conference colleagues, and that the lot of most lakes seemed to be worse.

WHO IS THIS BOOK FOR?

I wrote this book with the goal of reaching a broad readership, particularly people who are interested in natural history or lakes but are not professional scientists, or at least not specialists in ecology or evolution. This seems to me an important audience. When some precious patch of nature is about to be lost through human action—for example, if an old-growth forest is about to be logged—the brave individuals who sacrifice careers and livelihoods to protect it frequently have no specialized training. Sometimes a powerful personal commitment can be every bit as crucial as an advanced degree. Other nonspecialists make essential contributions by giving up meat, voting with nature in mind, or making donations to the conservation organizations that are working hard to keep more green in the world. And of course, it is nonscientists in positions of power who often make the most critical decisions about the future of nature.

More broadly, I hope I have written for anyone who is simply curious to learn a little more about the diversity of life. For myself, that reader is personified by my (late) father-in-law, Al Hodgins. Al was a high school English teacher who had long enjoyed a hike or a day of fly-fishing, but who only became an enthusiast of science and natural history writing when I started giving him works by Stephen Jay Gould. I still miss our conversations about those books. In a similar vein, one reviewer suggested that his parents would enjoy the chapter I asked him to read from this volume. I hope he is right, even if I have no expectation of rivaling Gould or Ed Yong. And I will be delighted if my efforts lead a few folks living along the shores of an ancient lake to feel more pride in their aquatic treasure.

I have also had in mind readers who are not quite professional scientists but not exactly laypeople either—that is, students. Really, it is younger versions of myself that I am thinking of. Accessible writing about nature and science was tremendously important to me during my childhood and youth, and I continue to enjoy such writing today. I still have copies of Herbert S. Zim's *Zoology* and similar Golden Guides from my childhood, and have lost track of how many times I have read Gerald Durrell's *My Family and Other Animals*. As a young adolescent I reveled in memoirs of field biologists like Eugenie Clark, dreaming of someday doing such work. Even as a university student majoring in zoology, I learned a great deal from the semipopular writings of E. O. Wilson, Gould, Dawkins, and others. I also found inspiration in their words, which may have been just as significant as any knowledge they transmitted. I hope a few nascent scientists, teachers, naturalists, and activists can find something they value in these pages.

My expert colleagues are an audience I cannot help but imagine looking over my shoulder as I write and revise, since they are the readers I know best and the ones most likely to catch a slip. Their imagined gaze can be especially intimidating when I am trying to synthesize and communicate topics on which I do not work directly, which is often the case in a book this broad. I have tried hard to be accurate, but also to resist getting bogged down in the qualifications and details that are sirens for scientists. Hopefully the balance I have struck seems sensible and any errors are minor.

There are enough lakes and intriguing studies that writing a truly comprehensive book would almost certainly mean writing a very long, repetitive one, so I have had to pick and choose which examples to present. I apologize sincerely to the authors of the many superb investigations that I could not fit into these pages, and acknowledge the biases that are inevitable given the limits of one person's knowledge. As well, I should note that I have not attempted to write a comprehensive review of how living things diversify or any other conceptual topic. Instead, my goal has been to convey the contributions being made by research in ancient lakes.

READING THIS BOOK

This volume was written pretty much as I imagined it would be read, from chapter 1 through 9. Still, I have tried to avoid having the later chapters rely too much on earlier material. Thus, the reader in a rush could likely skip a few stretches without getting hopelessly confused. The glossary might help too. Some chapters could potentially be read on their own, especially chapter 5 on some of the ways variation is maintained within populations, and chapter 8 on Baikal. But, of course, I hope most readers will find the whole book of interest.

I have provided a brief summary at the end of each chapter because I find these handy when I am the reader. Particularly if one's time with a book is often interrupted, it can be helpful to be reminded at the end of a chapter about its beginning, which one may have read a good while earlier. A brief summary can also ease the transition to the following chapter during reading.

Rather than scatter citations throughout the text, I have gone with the less formal arrangement of providing citations and suggestions for further reading at the end, organized by chapter and then topic. My intention is thereby to minimize interruptions; distractions are a bane of modern life, and few of us wish more. It is a somewhat informal, less academic format, but that seems all right.

CONSERVATION, CELEBRATION

When I started graduate school, most students of ecology and evolutionary biology received a training in basic research and expected to pursue such work throughout their careers, with the exception of those who chose applied areas like fisheries or forestry. Conservation biology was an emerging field that was still becoming established. Today the situation is different. Conservation-related questions and considerations play a role in the work of most researchers who conduct field studies or analyze data from natural systems. For many, investigating how natural populations and systems are responding to human-caused environmental changes is the central task.

One could easily devote a whole volume to the dangers facing ancient lakes and the ways in which they are already degraded. Indeed, it is tremendously important that we allocate much more time and resources to these problems. I devote the final chapter of this volume exclusively to those issues, and they come up elsewhere in the book as well. But my main goal is to celebrate the lakes and their life, for themselves, the wonder they inspire, and what they are teaching us about nature's fundamental processes. We need to better care for the marvelous living things uniquely present on our singularly green planet, but it is good to simply delight in them too.

1 WHY ANCIENT LAKES DESERVE OUR ATTENTION, AND HOW THEY GOT MINE

Teto'omo rilipu doro / Kadolidinya Rano Poso
Maramba kojo nasindi ando / Mampalindo raya mawo

Busy village / The beauty of Lake Poso
Beautiful to behold by the sun / There is peace in the longing heart
—"Kadolidi Rano Poso," traditional song of the Poso region, in the Pamona language

I was an eighteen-year-old zoology student when I arrived on the shores of Lake Matano, excited beyond measure to be on the Indonesian island of Sulawesi and seeing tropical biodiversity for myself. I was especially keen to visit the forests and had even arranged to collect insect specimens for the University of British Columbia, where I had recently begun studying. The lake itself was magnificent, clean and clear and fringed by verdant forest, and I asked both locals and expatriates if the lake's fish were interesting. Everyone I met assured me they were not. I would be much better off to try to get down to the coast to see the coral reefs, which I would be sure to love. This made sense to me as many hours of television nature programs, especially

Jacques Cousteau's, had long since convinced me of the wonders of corals. But what I had heard about Lake Matano was wrong.

The popular media notwithstanding, there is much more to biodiversity than tropical forests and coral reefs, marvelous as such systems are. In fact, the Malili Lakes of Central Sulawesi (especially Lakes Matano and Towuti), together with a collection of similar lakes scattered about the globe, contain biodiversity that is remarkable in extent, beauty, and uniqueness—and studies of these lakes are changing the way we think about how new species form and how fast.

I found myself at Matano because in the late 1970s, two of my uncles, employees of the mining company INCO (also known as International Nickel), had moved from their homes in Thompson, Canada, to Indonesia to work in a mine being started on Matano's shores; one of my aunts and a cousin also made the move and they all remained there several years. I could get to Sulawesi cheaply because my father was a pilot for a Canadian airline. As an aspiring biologist and naturalist, the trip was a natural for me, beyond the powerful pull of tolerant relations with a spare bed and full fridge.

The former Celebes is one of the world's most biologically idiosyncratic islands. It is famous in particular because of the work done there by Alfred Russel Wallace, the co-originator with Charles Darwin of the theory of natural selection. Wallace showed that Sulawesi marks a transition between the faunas characteristic of Asia to the west and Australia to the south and east. Like many before me, I was enchanted by the island and its natural history. It seemed a peaceful, if not quiet, place with forests full of buzzing, clicking, sometimes screeching insects

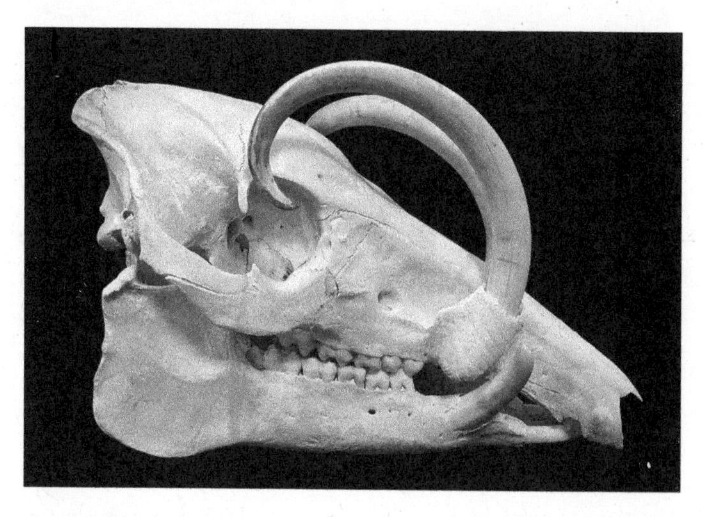

Figure 1.1
Babirusa skull, modified slightly from color original. *Source:* Didier Descouens, Wikimedia Commons. Published under a Creative Commons CC BY-SA 4.0 license (https://creativecommons.org/licenses/by-sa/4.0/deed.en).

among the tree ferns, pitcher plants, and other exotic vegetation. With a little luck, you occasionally got a glimpse of one of the various species of macaques found only on the island or even a babirusa—Sulawesi's peculiar "deer-pig" in which males possess canines that curve up and then inward, sometimes piercing their flesh (figure 1.1). Easier to see were sailfin lizards, which when young will run on water like the famous basilisks (or "Jesus lizards") of South and Central America, or white-bellied sea eagles, majestic creatures whose white and gray plumage was hard to miss as they soared regally over Lake Matano.

It was only years later that I learned from *The Ecology of Sulawesi*, a volume published well after my trip, that Matano contains a fascinating, if little studied, set of fish species. In addition, a second, almost completely distinct set is found just a few miles downstream in a sister lake, Towuti. When I read a series of papers on the Malili Lakes fishes published in 1990 and 1991 by Swiss scientist Maurice Kottelat, I became convinced that I must return to the lakes. I was most excited by Kottelat's descriptions of the *radiations*, to use evolutionary biology's term for diversifying groups, of the Telmatherinidae, or sail-fin silversides. But getting back took me until 2000, almost twenty years.

Fortunately, it was well worth the wait and effort; my return to Lake Matano was one of the most wondrous and inspirational experiences I have had as a biologist. Simply getting the permits to do the work proved a huge job, so receiving the approvals was the first exciting milestone. But once I got there—what fish! Their colors were brilliant, and their behavior frenetic, complex, and bubbling with research possibilities. Within a day I had made the observations that would propel a major branch of my research program for years. I saw striking variation in the colors of males of the same species—hues from extremes of violet through vibrant yellows and shades from near black to almost luminous white—as well as intriguing differences in the visual environments of different lake habitats. I could not help but wonder if variation in fish color patterns and light environments might be linked, and this was to become one of our key research topics.

The fish seemed to be spawning much of the time, but not the simple pairings typical of mammals such as ourselves. Instead, there were often two, three, or even four males sidling up to a female as she pressed her belly to the mud to release

Figure 1.2
Left to right: Fadly Tantu, Suzanne Gray, and the author on Lake Matano in 2003.

eggs, trying to give their sperm a few more lottery tickets for the next generation. Then it really got macabre as once the spawning ended, some of the fish spun around to bite at the muck in apparent cannibalism of their own offspring. Theoretical explanations for some of these behaviors occurred to me, yet it would require years of work before we would know if the evidence from this system matched the theory.

I also saw the variation in body shape and size, and in feeding behaviors, that had been described by Kottelat. Most obviously, some fish appeared to spend their time foraging in the *aufwuchs*, as German scientists call the layer of algae and tiny organisms that often carpets submerged rocks and hard sediments, while others followed courting pairs of fish around, brazenly darting in to search for eggs every time a spawning occurred. I also caught occasional glimpses of a rapidly moving, streamlined fish that I suspected was a species thought possibly to eat the scales off other telmatherinids, a peculiar and unsavory way of making a living that has arisen in several lineages on different continents (although the Matano species has turned out to be a more conventional predator). In the last two decades, my collaborator Fabian Herder and his laboratory have made a good deal of progress working out natural selection's role in the diversification of the telmatherinid fishes in morphology, feeding habits, and other traits in addition to helping resolve the history of these radiations. From these and other studies it is becoming clear that the formation of new species (speciation) is often tightly intertwined with ecological opportunity and can occur even within a single lake.

Fabian, myself, and various others were drawn to the Malili Lakes not just by their remarkable biodiversity but also the fact that they are not *overwhelmingly* diverse—at least not in comparison to the most famous lake radiations, the cichlids of the African Great Lakes of Tanganyika, Malawi, and Victoria (and some smaller nearby water bodies as well). Cichlids are a diverse group of colorful fishes, typically about five to twenty centimeters long, that are popular with aquarists. Each of these lakes contains 250 to 500 or more unique species of cichlid fish. They provide some of the most iconic examples of adaptive radiation, which can be more formally defined as the rapid diversification of an ancestor into various new species through natural selection arising from different environments or resources.

To convey the scale of these radiations, let's consider Lake Malawi, with at least 500 cichlid species (possibly 850). I say at least because the lake's cichlid diversity has not been fully described, there is disagreement about when to call two forms different species, and extinctions are likely occurring. Nevertheless, this single body of water without question has more species of cichlid in it—almost all evolved right there, in that lake—than the entire United States has mammals, reptiles, or amphibians. A more like-for-like comparison is with the vast but young inland sea of Lake Superior, the largest of North America's Great Lakes. About 10,000 years in age, Superior is home to less than 100 fish species, many of which arrived recently with the intentional or unintentional help of people. Most of its fauna is also found in other lakes.

In a different African Great Lake, Victoria, the number of cichlid species is again immense and the pace of speciation is even more extreme, with a faster sustained pace than in possibly any other vertebrate lineage. One can argue, though, that while cichlids and the African Great Lakes are special, we should expect them to be at least a bit exceptional merely by virtue of being in fresh water. It is not widely appreciated, regrettably, that freshwater habitats are unusually diverse almost across the board. Despite covering less than 1 percent of the planet's surface, fresh waters harbor more than half of all fish species (for at least some of their life cycle) and one-quarter of all vertebrates. The ratios are still more extreme when framed in terms of the volume of fresh water relative to salt water. Even so, Africa's Great Lakes stand out conspicuously.

Beyond impressive numbers of species, the cichlid radiations of the African Great Lakes have resulted in some genuinely odd creatures. Among my favorites is Tanganyika's *Neolamprologus pulcher*, which has been studied in both field and lab by

Barbara and Michael Taborsky and their colleagues at the University of Bern. As in wolves, scrub jays, bee-eater birds, and a small set of other vertebrates, members of this species breed cooperatively. Some adults forgo rearing their own families to help raise the offspring of others, which may mean caring for brothers and sisters, though not always. From the perspective of conventional Darwinian natural selection, such altruism is puzzling, and heated controversies over how to explain it continue to smolder in journals. In retrospect, I suppose I should not have been so surprised to learn while a graduate student that quixotic social arrangements could be found in freshwater fish; yet years later, I still find this species remarkable.

Another oddity, and one with broad evolutionary implications, is found in the sex determination systems of African cichlids. Most of us don't think too much about the genetics of the sexes, taking it for granted that the presence of a Y chromosome makes one genetically male, and that is about it, although of course in our species one's genetic sex and one's self-identification/perception may not be the same. But since X-Y is how sex is determined not just in ourselves but also in most of the animals we interact with day-to-day, such as dogs, cats, and most farm animals—indeed almost all mammals—the matter might seem settled and inflexible at the chromosome level. It is not. In birds and butterflies, for example, it is the presence of a W chromosome that makes a female, while males have two Zs. The Zs are analogous to the two Xs of human females, but with the sexes reversed. Still, birds and butterflies are pretty big, old groups comprising thousands of species each, so even after taking them into account, sex determination still looks predictable and orderly. In African cichlids, this falls apart. In cichlids, multiple different genes may code for being male or female in utterly different ways, and on different chromosomes within a single species.

In light of theories of sex determination and sex chromosome evolution, this is less surprising than one might guess. Theory predicts that if an allele (an allele is a form of a gene—i.e., different alleles code for blue or brown eyes in humans) has benefits in one sex and costs in the other, it will be favored more strongly if it occurs more often in the sex it helps. So an allele that makes sperm swim fast when in males but disrupts ovulation when in females will be most successful if it is frequently present in males and rarely in females—for example, if it sits on the same chromosome as and near to an allele that codes for being male. This is roughly equivalent in a mammal to being on the Y chromosome. In some Lake Malawi cichlids, an allele for a color pattern that appears advantageous only in females is tightly linked to an allele (not yet identified) that makes fish female. Moreover, this allele makes a fish female regardless of which X or Y chromosomes are present—the familiar XY sex determination system is also found in these and many other cichlids, but it can be overridden. As if that were not enough, a similar color pattern is present in Lake Victoria cichlids and associated with sex, but it appears to result from a different allele than in the Malawi species.

Given their apparently distinct genetic foundations, the female color pattern and sex determination systems in Malawi and Victoria are an example of convergent evolution, defined as the independent evolution of similar traits in different lineages. Convergence is important in part because when traits are consistently associated with specific environmental challenges or selection pressures across lineages, some of the most compelling evidence for evolution by natural selection is obtained. Convergent evolution is rampant in the cichlids and other ancient lake groups for traits ranging from sex determination and color pattern, as in this case, to tooth shape, body shape, color vision, and more.

With new genomic methods and the rising flood of genomic data, we are no longer confined to describing convergent traits at the level of what can be seen or directly measured. Now that we have sequenced all the DNA for thousands of species, and have the ability to track sequences across generations and lineages as well as even edit genes, we can often work out a trait's molecular genetics. It is possible, for instance, to ask if the same or different genes are used to generate similar shapes and other features. One unexpected result of such studies is the growing number of examples in which similar traits from different species are indeed the result of changes in the same genes. Further, they sometimes involve not just the same gene but the very same mutation too; this finding was a major surprise—one that can be explained by interspecific matings allowing such mutations to move between species. With the proliferation of genomic tools, the evidence is mounting that such hybridization and the exchange of genes between seemingly distinct species is much more common than had been thought. It has also been suggested, and now supported by increasing evidence, that widespread hybridization in the early stages of adaptive radiations may be a key catalyst to rapid diversification. This idea is largely due to Ole Seehausen of the University of Bern—one of the most creative researchers working in ancient lakes.

The breathtaking diversity in the numbers of species, feeding behavior, social behavior, sex determination, and numerous other features seen in the Malili Lakes and African Great Lakes inevitably prompts the following question: Is it commonplace for lakes to contain dozens or hundreds of unique species if we look hard enough?

Sadly, it is not. Many lakes contain distinctive populations of widespread species, such as lakes with threespine stickleback fish, a species that has become an important model for evolutionary biology and that I also study. Some lakes possess a single unique fish species or perhaps a pair of sticklebacks, but few host substantial radiations. One major reason is probably insufficient time. Most lakes are young, less than 10,000 years of age, and unlikely to persist a great deal longer. This is because many lakes soon fill with sediment, or in polar regions, are buried under ice when the glaciers come back. And a great many lakes, especially in northern nations like my original home of Canada, only formed as the last round of glaciation ended about 12,000 years back.

Lakes like Matano and Malawi, relatively old and with substantial numbers of unique species, have been referred to as "ancient lakes" since at least 1950, when John Langdon Brooks published an influential synthesis titled "Speciation in Ancient Lakes." Brooks surveyed four additional lakes beyond those we have already considered, and since that time a consensus has emerged that more should be added to the category. There is no universally agreed-upon definition, however, and thus no indisputable list (the lakes that will feature most often in this book are shown in figure 1.3). A recent overview of ecological changes in ancient lakes suggests they be defined as lakes that

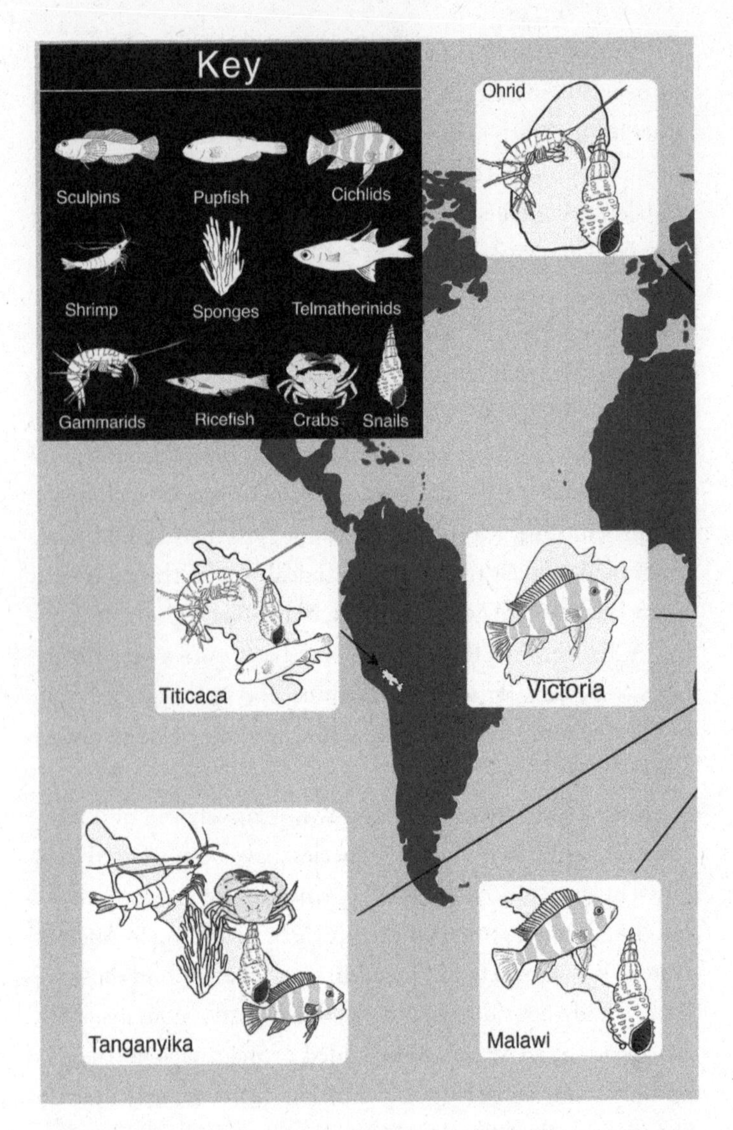

Figure 1.3

The best-known ancient lakes and the main radiations featured in this book. Lake details are approximate, for illustrative purposes, and not all radiations (e.g., ostracod crustaceans and diatoms) are illustrated. *Sources:* Haleigh Mooring; modeled in part after Cristescu et al., "Ancient Lakes Revisited: From the Ecology to the Genetics of Speciation," *Molecular Ecology* (2010).

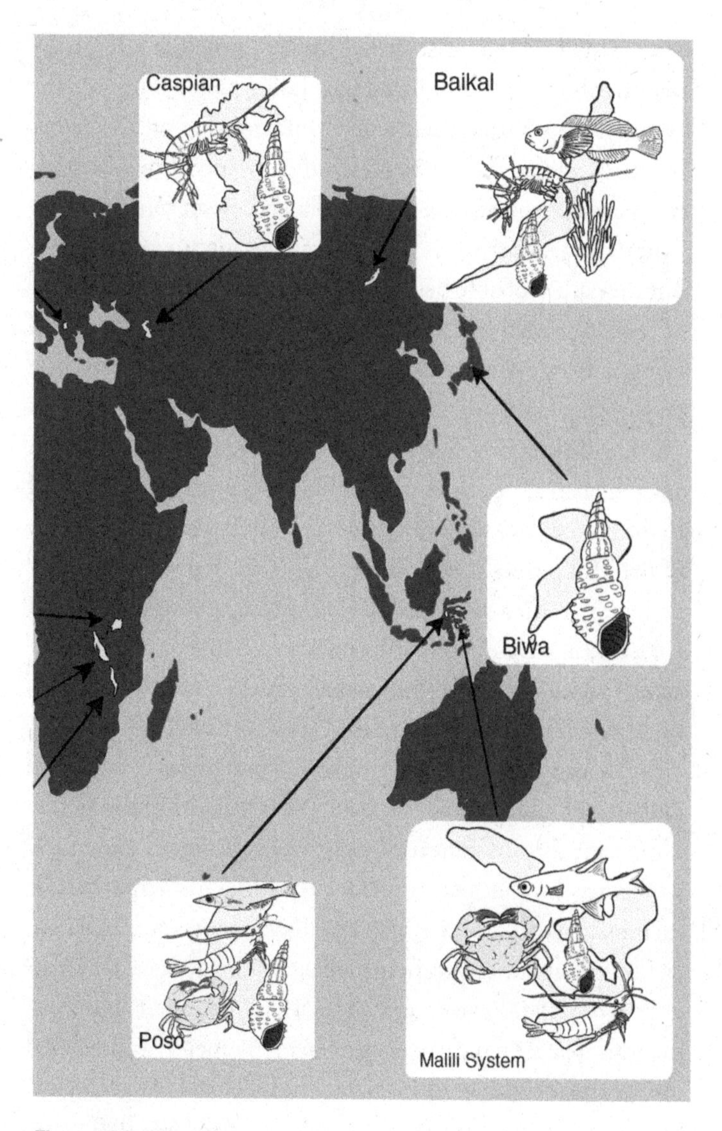

Figure 1.3 (continued)

have existed since at least the beginning of the last interglacial period, about 130,000 years ago. Using this definition, the authors, who started their work while meeting on the shores of Japan's ancient Lake Biwa, identified twenty-nine matching water bodies spread across Australia and every continent but Antarctica (Antarctica is home to fascinating subglacial lakes, but they are probably best for another book). The consistency of their approach (see figure P.1) is attractive. Yet one effect of using a strictly time-based classification is that not every lake on their list has many unique species in it.

I will discuss a few less biodiverse and/or well-studied ancient lakes at points in this volume, in particular with regard to the tragedies of the Aral Sea of Central Asia and Lake Lanao of the Philippines. Other intriguing lakes that will hopefully be better studied soon include Ohrid's neighbor Prespa and Lake Inle of Myanmar. North America is home to four ancient lakes, including the well-known Lake Tahoe, but none host distinctive faunas on the scale of the lakes illustrated in figure 1.3.

The observation that not every old lake hosts extensive radiations highlights the fact that time is just one of the factors important for diversification and the evolution of endemics, which are unique species present only in a limited area. For example, Lake Victoria, which certainly harbors more endemic cichlids than does Lake Tanganyika, appears to have dried up only about 15,000 years ago—a victim of the water level fluctuations that are typical of ancient lakes (frequently young lakes too) as a result of changes in climate and geology. The geological basin in which Victoria resides is older, perhaps 400,000 years, and because the whole basin did not dry out, Victoria's entire fauna did not go extinct during that drying. But no matter how you measure age, Victoria is both younger and hosts more endemic species than Lake Tanganyika, which is approximately ten million years in age. Some would argue that Victoria should not be included in a discussion of ancient lakes, but Victoria's cichlids possess an unusual form of antiquity that will emerge as their story unfolds, so I will include Victoria. Working out why only some lineages go through extraordinary radiations and lakes of a similar age host different numbers of unique species are two of the major challenges for those studying ancient lake biodiversity.

So far, I have described tropical ancient lakes, and in general there is a great deal more biodiversity in tropical regions than in temperate or polar areas. This remains true whether one is looking at terrestrial, marine, or freshwater habitats. But perhaps the most remarkable of ancient lakes, the one Brooks began his 1950 paper with, is just about as untropical as most of us want to think about: Lake Baikal of Siberia.

Baikal is impressive in almost every possible way. At twenty-five million years of age (or more), it is the oldest lake. And its statistics for size are also extraordinary: Baikal is 636 kilometers long, almost 80 kilometers in width at its broadest point, and reaches 1,642 meters in depth, or the deepest of any lake on our planet. Put together, these numbers yield a volume much greater than that of any other freshwater lake, with a surface area also in the top ten. What is most unusual about Baikal's depths, though, is not so much their extent as what they contain. Many lakes are stratified at least some of the year, divided by depth into layers with different temperatures and sometimes different concentrations of dissolved gases or other chemicals. This can result in little oxygen in deep water and consequently little life, especially large organisms. Baikal, however, has relatively high oxygen levels at depth and a distinctive deepwater fauna.

Among the lineages that have colonized Baikal's depths is one of its best studied, the amphipods—small, laterally compressed crustaceans—of the genus *Gammarus* (figure 1.3) and related genera, which have diversified into over 265 species. In Baikal they have achieved a diversity of sizes otherwise seen only in the oceans and an even more extensive diversity of forms. These creatures, which typically look like diminutive shrimp, will be familiar to many who have used them as fish bait (scuds), or encountered them during a biology or limnology class that involved scooping and examining tiny creatures from the bottom of a pond. I have a perhaps unreasonable fondness for them myself, having spent many childhood hours watching them scoot around pickle jars after I caught them from the drainage ditch in front of my family's home (I think ditches are an underappreciated incubator of future biologists). They have an intriguing mate-guarding behavior in which males typically grab onto females and don't let go until the female molts, at which point her eggs can be fertilized. Such behaviors along with a general hardiness also make them well suited to introductory biology laboratories.

The amphipods of Baikal include some run-of-the-mill *Gammarus* much like those seen in many freshwater and marine locales, feeding on detritus along the bottom. More interesting are the species that have taken up novel ecological roles, including predatory forms, parasitic types that make their living from other amphipods, and one highly unusual species that lives in the open water, migrating from the depths to shallower waters every evening to feed. This radiation includes "armored" species reminiscent of medieval knights and exceptionally large gammarids, leviathans of the freshwater amphipod world that can reach 9 centimeters in length.

In light of the diversity of forms and ecologies in this radiation, it was long thought that the lake must have been invaded repeatedly by different lineages of amphipods, and as many as nineteen such invasions were hypothesized. More recent molecular studies suggest fewer such events, but this is an active area of research.

Other members of Baikal's fauna are reminiscent of marine forms and contributed to early speculations that some ancient lakes were former seas. In particular, Baikal's sponges are not the drab, beige creatures often encountered in fresh water but instead branching structures of 1.2 meters or more in height that can form "forests" along the bottom of some parts of the lake. Their color is also striking, with some species an attractive green as a result of a photosynthesizing microbe that lives in their tissues. Even more evocative of the ocean is the "nerpa," confined entirely to Baikal and the only species of seal that lives exclusively in fresh water. I was lucky to see Baikal's sponges when I visited there for a conference, and they were impressive, but despite spending time on the lake and looking for them, I did not see any nerpa. Fortunately, this is not because they are disappearing. They number in the tens of thousands despite centuries of exploitation, the stress of serious pollution, and major disease outbreaks; it is refreshing to find a creature thriving that one might expect to be in trouble.

Baikal has persisted for so long because it is not the result of short-term processes such as glacial advances and retreats

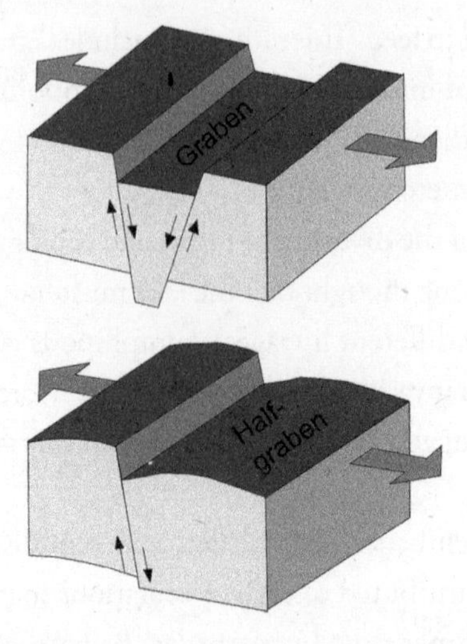

Figure 1.4

Grabens and half-grabens, the geological formations underlying many ancient lakes. *Source:* Modified from Aymath2, Wikimedia Commons. Published under a Creative Commons CC BY 3.0 license (https://creativecommons.org/licenses/by-sa/3.0/deed.en).

along with their associated flooding and drying. Rather, it is a product of tectonic activity involving changes in the earth's crust. Baikal is more specifically a tectonic graben—a lake produced when a long, narrow block sinks between two parallel faults and the basin fills with water (figure 1.4). Over half the ancient lakes in figure 1.3 are grabens or related half-grabens, and all of them are tectonic in origin. Of the freshwater lakes in figure 1.3, Baikal and Tanganyika are by far the oldest, and the largest too, by volume. Malawi is next in size, followed by Victoria and then Titicaca. The Caspian Sea has the greatest area, but is not fresh water, though with only about one-third the salt of the oceans on average. It is considered a lake because it has no connection to the ocean.

Deep as Baikal is, it would be much deeper but for the kilometers of silt that have accumulated beneath it as the millennia have rolled by. Happily for evolutionary biologists and ecologists, deep sediments are present in many lakes and provide an often-continuous record, in layers conveniently arranged ancient to recent from bottom to top, of conditions in a lake, what lived there, and what lived around it. Ecologists interested in what is happening on land frequently identify pollen in the layers of silt extracted from sediment samples and thereby identify the land-dwelling plants of times past, also enabling inferences about past environments.

Much more can be done with lake sediment samples than identifying land plants, of course, including addressing a surprising shortcoming of research on the origins of modern biodiversity: the often-platonic relationship between evolutionary studies of living organisms and studies of fossils. Comparative biologists will sometimes use fossils to calibrate the evolutionary trees they are building and work out rates of change, but the relationship rarely goes much further, at least in any systematic way. Yet in lakes, especially old ones, aquatic biologists with access to sediment records can readily look, quantitatively, at the hard parts of aquatic plants and animals to infer what was present. Moreover, with improvements in the analysis of ancient as well as environmental DNA (which has been released from an organism into the environment), and the collection of samples that are being handled with these analyses in mind, extraordinary new insights are possible. The promise of these more evolution-focused studies of lake cores is at the earliest stages of being realized, but interesting results are starting to emerge. For example, new samples from Albania and North

Macedonia's Lake Ohrid, which is at least 1.3 million years of age, are being used to evaluate the relative importance of environmental changes in propelling speciation in the lake.

Studies of lake sediments and genomes not only can provide powerful insights into evolutionary history but help us to anticipate the future too, specifically the effects of coming environmental changes. It is thus a curious and useful quirk of ancient lakes that their age, and the long history embedded and encoded in their slowly decreasing depths, are improving our focus as we attempt to look forward so as to plan for and manage a future in which our greatest challenges may be ecological.

* * *

When we think about biodiversity, we often think of habitats such as tropical forests and coral reefs. Yet a disproportionate amount of biodiversity is in fresh water, and many unique species are concentrated in tectonic lakes whose origins precede the last round of glaciation—ancient lakes. They include Baikal, Titicaca, the African Great Lakes, and additional lakes scattered across the globe. They are home to numerous evolutionary oddities and some of the most extreme diversifications known, and their study is changing how we think about the formation of new species and how life diversifies.

2 THE ECOLOGICAL CAUSES OF DIVERSITY

shihō yori / hana fuki irete / nio no nami

From all four quarters / cherry petals blowing in. / To Biwa's waters!

—Basho, 1690

We seldom hear about chimprillas or gorpanzees. If you are lucky enough to find yourself in a forest in Central Africa—and happily there are such forests quite close to lakes I have visited—you might, with a little luck or effort, see chimpanzees. With some more effort you might see gorillas. Either way, you will not come across any creatures that are halfway in between, whether for ear shape, brain size, skull shape, testis size, or any of the long list of characteristics that separate these relatives of ours. And you certainly won't happen on a population of apes in which each individual possesses a haphazardly assembled collection of chimpanzee and gorilla features and traits. Rather, you will encounter members of distinct species, and where they occur side by side it will not be hard to distinguish one from the other, even if the odd chimp is big enough to be mistaken initially for a gorilla.

As most of us know intuitively, this is what we usually find in nature, especially in the sorts of animals and plants that can be seen with the naked eye. At any particular location, the diversity of life is separated into pretty much discrete packets that are known as species. Each species comprises populations of individuals that share many traits with each other, but differ in a variety of ways from members of other species. There are exceptions of course (there almost always are in living systems), but for most organisms in most places, the great majority of individuals are readily assigned to one species or another. This is why field guides are useful, and why they are profitable enough that they keep being published. There has been a modest amount of argument among biologists about the reality of species, with the opposing view being that they are arbitrary constructs of the human mind, but most of us find the concept useful and rely on it in our work.

Still, perhaps this is not how things had to be. Why *is* life organized into these more or less nonoverlapping units? How exactly do we decide when we have two species rather than one? And are there better ways to describe biological diversity than just counting species?

Let's start with a definition. Unfortunately, coming up with a species definition that a good number of biologists can agree on turns out to be no small thing, and several are in use. Even philosophers have gotten in on the act and generated a surprising volume of verbiage concerning what a species is from a philosophical perspective. But a great many biologists, and especially those whose research focuses on how species are formed, mainly rely on a single definition. It states that a species is a set of populations (or members of a single large population) that actually or potentially interbreed and exchange genes only with each other; this is known as the biological species concept. Oddly enough, it is not something that Darwin handed down to us, at least not explicitly. Although his most famous work was titled *On the Origin of Species by Means of Natural Selection*, Darwin for the most part treated species as extremes of the varying forms he saw throughout nature. He did not give much special attention to this unit in the *Origin*—though he did comment on it here and there, and considerably more in some of his unpublished writings.

The biological species concept was due mainly to Ernst Mayr, a remarkable scientist with whom I briefly overlapped, to my good fortune, while starting my graduate studies at Harvard in the late 1980s (there is a photo of him in chapter 4). Professor Mayr was then in his own eighties and still publishing regularly—something he carried on doing almost continuously until he died at age 100. I well recall his aggressive questions and comments at seminars, and he is reported to have remarked, "I'm not dogmatic, I'm simply right!" We students held him in awe.

The main problem with Mayr's "BSC," as it is often known, arises when dealing with sets of populations that have wide distributions. Things really go downhill when there are breaks in distributions. In the Malili Lakes of Sulawesi, for example, there is a ricefish (*Oryzias marmoratus*) that is found in more than one lake and varies somewhat between lakes. Since the lakes are separated by rivers that are likely more difficult for the fish to traverse, movement between lakes and opportunities for interbreeding may be rare. Would fish from different lakes interbreed if placed together, and would their offspring survive and themselves be able to breed? Maybe, but with so many millions of species of animals and plants we can rarely do the necessary tests to address these questions—certainly not for every break in the distribution of every species—and it is often necessary to make a judgment call (highlighting these challenges is the fact that the evolution of *O. marmoratus* has become an increasingly complex matter just during the writing of this book). With more molecular data now available, we frequently have a better idea just how genetically distinct different populations are, but judgment is still involved in deciding when we have two or more species rather than just one.

Distinct species that live alongside one another yet occasionally interbreed can also complicate life for the biologist trying to apply the BSC. It can be tough to decide just how much interbreeding is too much and at which point one species should be treated as two. In addition, the BSC only works when reproduction is sexual and thus interbreeding is a possibility; no sex means no interbreeding and an irrelevant BSC. This was long perceived to be a major problem for many microorganisms, especially bacteria. Lately, however, microbiologists are finding that exchange of genes with similar individuals, much like what happens in sexual reproduction, is more common than once thought for a wide range of microscopic creatures. Hence the BSC, its difficulties notwithstanding, may apply even more broadly than once supposed. In fact, it is now clear that even viruses can exchange genetic material when different strains infect a single host at the same time. Moreover, they exchange genetic material much more often with genetically similar viruses, contributing to clusters of individuals much alike in genes and form, and surprisingly comparable to the clusters that comprise an animal species. All in all, then, the BSC remains the species definition of choice for most, but those who study speciation appreciate its limitations and are increasingly bypassing them by treating speciation as a process that leads to a continuum of levels of reproductive isolation, rather than to necessarily discrete outcomes. Taxonomists, however, have to assign species names and are obliged to use their best judgment in this sometimes arduous task.

IS BIODIVERSITY THE NUMBER OF SPECIES?

Well-defined species concepts are important in the study of biodiversity in part because they enable us to delineate species in a reasonably consistent way and count the number present in a lake, or whatever sort of unit one is interested in. We can then compile counts for different lakes as one way to compare biodiversity across localities, or we can use slightly more complicated indexes that are based on relative abundance as well as species counts. For instance, a lake with five clam species in it, all equally abundant at 20 percent of the total numbers each, might be considered more diverse than a different five-species lake where 98 percent of the clams are of one type, with few of each of the other four. There are also some quite different approaches that can be used to quantify the biodiversity of a lake. One important one, which is gaining in popularity for conservation work in particular, is based on calculations of how much unique evolutionary history is represented by the organisms in the lake.

Basically, the patterns on an evolutionary tree for a group of creatures are used as the basis for calculating how much unique history is present in any one species. This can be note-worthy for its own sake and can also be a guide to how unusual the organism might be for characteristics of interest but not yet examined—say, its immune system if biomedical issues are the focus. An evolutionary tree, known to biologists as a phylogeny, shows the relationships among a group of species as a branching pattern extending from an ancient ancestor to modern descendants, with intermediate ancestors at the branching points. The longer a branch (in time, DNA sequence differences, or more often both) leading to a species and its near relatives, and the fewer near relatives it has, the more unique history it possesses (figure 2.1).

Evolutionary trees are roughly analogous to family trees (genealogies), so here a family tree can help illustrate the idea. Suppose there were two sisters, each with children. Sister Alice had five children while sister Viola had one. Each of Alice's children also had five offspring, who in turn had five each. By the third generation of her descendants, Alice has accumulated 125 great-grandchildren (this is actually a bit like my Grandmother Thompson's family from rural Canada, but I digress). Meanwhile, Viola's only child kept up the family tradition and also had an only child, who did the same. After three generations, Viola has just one descendant, Victor. Victor is the sole receptacle for three generations of his family's genetic history. He is pretty unique, even if his family reunion picnics are a bit quiet and lonely. In the meantime, over on Alice's branch of the family tree, her great-grandchild Andrew goes to huge picnics. He has plenty of company from folks who are a lot like him, and you could hardly say there is much unique about Andrew—with four siblings, twenty first cousins, and 100 more distant

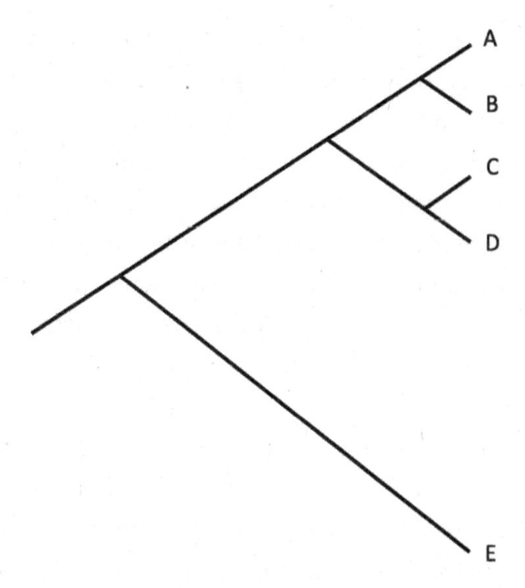

Figure 2.1
Evolutionary tree, or phylogeny, on which modern species "E," alone at the end of a long branch, has more unique evolutionary history, shared with no other species, than does species "A," or any of the other species shown. The past is on the left, and the present is on the right.

cousins (not to mention all the uncles and aunts, great-aunts, and so on). If you were charged with choosing the person who carried in their genes the most unique history, you would have to go with Victor over Andrew. Lose Victor and a long tree branch is gone; lose Andrew and a small twig is lost from a bush.

For an evolutionary example, we can compare the ricefish lineage in Lake Matano, Sulawesi, to the lineage of sailfin silversides found there. Ricefish, mentioned briefly earlier, are a group of small fish—the size you would come across in a pet shop—that are found in East and Southeast Asia. There are about thirty-six species in total, with more on Sulawesi than anywhere else. They have become important as a model system for basic research; the medaka, *Oryzias latipes*, a Japanese ricefish, is used for a wide variety of biomedical investigations. There is just one ricefish in Lake Matano, which we will assume for our illustration has been evolving there on its own for about as long as Matano has been a lake—something on the order of one to two million years. We will assume the sailfin silversides started evolving in Lake Matano at about the same time, but were more prolific, with about ten silverside species now living in the lake. So if you were to consider only the fish in the lake and ask which fish has a more unique evolutionary history, the ricefish or Matano silverside, *Telmatherina prognatha*, you would probably have to go with the ricefish. This sort of approach can be used as the basis for prioritizing areas to protect or designing sets of protected areas that conserve as much evolutionary history as possible. Some folks working in conservation and management are now grinding through these sorts of calculations. Usually they take into account what the risk of extinction is for each species as well. For now, such sophisticated approaches may not be practical for the ancient lakes covered in this volume. But as management of the lakes gets more intense and evolutionary data accumulate—which is happening quickly—this could change.

The third approach you might take to quantifying biodiversity, the last one we will consider, would be to use form and function. This approach requires more data, but sometimes it is possible. Rather than just counting species, or calculating how distinct one or more species are in terms of their location on a family tree, one can calculate how distinct a species is (or set of species are) for one or more traits. Thus, a fish with an unusual diet, for example, or that is exceptionally big or tiny in size, might be considered more distinct from a functional perspective than one possessing a common diet or that is of an average size for its lineage. Thinking back to Alice's large family and Viola's modest one, suppose Alice's great grandson Andrew was 7 feet (213 cm) tall and everyone else on either side of the family tree was around 5'10" (178 cm) plus or minus a few inches—including Viola's one great-grandchild, Victor. If we were focused on distinctiveness in terms of height rather than family history, we might now reverse our original ranking and decide Andrew was more noteworthy than Victor.

This approach can be applied to our fish example. Our Matano ricefish seems so far to be a fairly typical ricefish—with nothing especially distinctive about it. *Telmatherina prognatha*, however, is a bit unusual for a lake fish in that as an ecological specialist that preys on other fish, it has evolved a torpedo-shaped body and wide jaw gape. This is distinctive because most fish species in freshwater lakes, especially from its family, eat invertebrates. So despite the fact that it has several close relatives and does not possess as much unique evolutionary history as the ricefish, it is more distinctive in terms of its feeding ecology and the features it has evolved that help it make a living feeding extensively on other fish. Consequently, if our focus was ecological distinctiveness, we might rank *T. prognatha* above the Matano ricefish. And we might rank lakes that possess a higher number or proportion of ecologically unusual species more highly in terms of their biodiversity than those filled with more conventional creatures—even if those conventional creatures embody a lot of independent evolutionary history (features at the ecosystem level are sometimes also emphasized).

In practice, though, differences in form and function are often correlated with uniqueness in terms of tree position and branch length, so we don't have to choose between them. Moreover, tree measures are sometimes used as a proxy for form/function, when only limited information is available about the ecology and natural history of the creatures in question.

SPECIATION

Regardless of how you choose to measure biodiversity, to get very much of it requires speciation. Since we are used to thinking about biodiversity in terms of numbers of species, this might seem self-evident—but there is a more subtle and important way in which the formation of new species enhances biodiversity. A new species is an independently evolving unit. A pair of new species cease to exchange genes with each other (or nearly so); they are free to diverge, to evolve in new and different directions, with neither holding the other back. As speciation events and species accumulate, the opportunities for independent evolution and diversification multiply.

The barriers to gene exchange that separate species can take many different forms. Who mates with who is of course critical, and mating between incipient (i.e., not fully isolated, often young) species can slow down or stop for a variety of reasons. Sometimes these new species mate in different habitats or at different times of day or year (or even in different years in the case of some salmon and cicadas!), or reject potential partners of the other species when the opportunity to interbreed does arise. In other cases, they are physically unable to mate, for example, if their genitalia have become too different. Even if they mate, the eggs and sperm may never find each other, or fail to fuse and begin development. Should things get that far, hybrid offspring may still not develop properly or may be less likely to survive. And even if they survive, they may be sterile or fail to find mates. Things can break down in a lot of different ways.

How and why these barriers evolve are the central questions of research on speciation. In addition to providing us with an influential species definition, Mayr set the direction and tone for fifty or sixty years of research in this area when he argued that in order for incipient species to diverge enough to stop exchanging genes, they each need some time alone. Each would need to spend a period of time evolving independently as a result of a geographic barrier of some sort—say, a piece of land separating two lakes when the water levels fall (for a fish) or a river breaking up a piece of land when the water levels rise (for a squirrel). The key notion is that a temporary physical barrier to gene exchange is needed in order for a permanent biological barrier to have a chance to evolve. Thus, if an ancestral, river-dwelling ricefish colonized two different lakes that had no connection to each other, after some thousands or millions of years the ricefish in each lake may come to differ. They may diverge enough that if the lakes happen to become connected, the two populations do not exchange genes and can persist as distinct forms. They might mate at different depths, at different seasons, or have different courtship behaviors, or maybe they have developmental differences that cause their fertilized eggs to die before hatching. There may be a few different biological isolating barriers preventing the exchange of genes between them.

With Mayr and others arguing aggressively that speciation was nearly impossible without a period of physical isolation to jump-start things, ambitious young biologists were keen to challenge what some saw as a dogma. The problem, though, was proving a negative; it can be difficult to obtain definitive evidence showing that currently overlapping species were never separated in a past that we did not see and cannot visit. Even so, persuasive case studies were eventually documented to the satisfaction of most evolutionists, and some of the most compelling and widely discussed examples came from lakes. Researchers basically showed that the various fish species found only in a single lake were all more closely related to each other than they were to any fish outside the lake—and the shape and history of the lake made a past period of geographic separation seem unlikely. In the most famous cases, these are relatively young crater lakes, which possess simple bowl shapes because they are the remnants of extinct but geologically young volcanoes. One recent example of speciation in a bowl-shaped lake comes from a more ancient water body, Sulawesi's Lake Poso.

This work was conducted by a group of scientists based mainly at the University of the Ryukyus in Okinawa, toward the southern tip of Japan and not so far from Sulawesi, at least as distances go in the tropical Pacific. Nobu Sutra was the lead author on the paper, which also included K. W. Masengi of Sulawesi's Sam Ratulangi University. The team collected tissue samples and made measurements of body shapes from the three ricefish species found in Lake Poso as well as from the nearest populations that they could locate of other ricefish. Using a method that provides extensive samples of DNA sequence spread across every chromosome, they were able to show that the three ricefish species in the lake are indeed more closely related to each other than to any ricefish from outside the lake, at least for most of their DNA.

An ecological aspect to these speciation events was suggested by differences in shape and size between the ricefish. In addition, they found intriguing evidence for some curious complications to the main findings: their analyses suggested at least one and potentially three episodes of gene exchange occurred between nascent ricefish species, one possibly from a population outside the lake, and the best-documented episode between species within the lake. Such an exchange preceded at least one speciation event and possibly both. Thus, the geographic context of speciation was a little messy—a finding arising time and again as genetic data get better and better, and as our perspective on what is possible becomes a little more flexible. The time frame their analyses suggest is consistent with Lake Poso's estimated age of one to two million years, and Poso's shape would make geographic separation within the lake implausible. It would be helpful to sample more extensively within and around the lake in order to assess these conclusions more definitively, but right now it looks like these ricefish evolved into new species within the lake.

It is increasingly well accepted that speciation sometimes happens with extensive or complete geographic overlap, as in the Poso ricefish example, but the prevailing view among evolutionary biologists is that most speciation indeed involves periods of geographic isolation between incipient species, much as Mayr argued. Speciation in the presence of ongoing gene exchange may occur more frequently when there is only a small level of geographic overlap—say, between fish in a river and a lake.

A shift in speciation research took place in the 1990s, an exciting one. The focus moved from working out the geography of how species form, which was really about the extent to which gene exchange *prevented* speciation, to identifying the processes that *accelerated* species formation by enhancing the evolution of biological barriers to gene exchange. Thus, the shift was from studying what prevented speciation to working on what propelled it. Sounds more positive, doesn't it? Certainly it seemed so at the time, at least to me. Some of the most influential ideas in this area came from the laboratory of Dolph Schluter at the University of British Columbia, where I received valuable and inspiring training at the start of my career. One of Dolph's great insights was to recognize the critical role that ecology, through ecologically based natural selection, could play in causing the evolution of reproductive isolation. Equally important was working out the signature patterns that would help us recognize when natural selection had been the driving force rather than other, more haphazard processes.

I have mentioned natural selection a few times already, and it will be a central topic in this book, so a definition seems in order. Defining it properly also helps reveal the marvelous simplicity and inevitability of the process given a few quite ordinary conditions. Just as a piece of granite is bound to sink if you drop it in a lake, evolution by natural selection will occur if there is variation in a trait, that variation is heritable (the offspring resemble their parents), and one form of a trait consistently has higher reproductive success than others—that is, individuals with one form of a trait contribute more offspring to the next generation. To illustrate, suppose some fish have larger mouths and some have smaller ones; parents with larger mouths have offspring with larger mouths (i.e., there is a genetic component); the fish are eating large prey, and fish with larger mouths get more to eat, survive better, and leave more offspring. Each generation, the average mouth size in the population will increase. This is evolution by natural selection. It is that simple. If the conditions are met, selection and evolution happen, just as a rock must sink if the conditions are met of being in a less dense liquid and the presence of gravity. Given certain starting conditions, some outcomes almost *inevitably* come to pass.

In addition to his evolutionary insights, Dolph has a talent for presenting an idea or finding in a memorable, compelling way—an ability often present in the most successful scientists, especially today when the scientific literature is exploding and the competition for attention is fierce, not only attention from readers, but from reviewers of grant proposals. Scientific tools are expensive, and grant reviewers hold the keys to the vault.

As he was developing his ideas about natural selection's role in speciation, Dolph made informal presentations during our laboratory meetings, mainly to the graduate students and postdoctoral fellows. These are often the most exciting settings for science, and it was at one of them that I first heard the term *ecological speciation*. Dolph used it to describe the evolution of barriers to gene exchange that arise, in a diversifying lineage, as a result of contrasting adaptations to different ecological conditions. At the time, numerous types of speciation had already been proposed and the literature was muddied by the sometimes-confusing jargon that accompanied them. I was doubtful about the value of encouraging another bit of specialized speciation verbiage and said so at one of the meetings. But

Dolph knew that this case was different; the term captured a valuable insight and efficiently conveyed an important concept. He also emphasized its connection to natural selection rather than using it more loosely to reference any of the ways by which ecological differences might influence the speciation process.

A straightforward illustration of ecological speciation is when a fish comes to occupy new ecological niches as it colonizes a lake that previously lacked fish, causing one species to divide into two or more. For example, in Lake Matano, Sulawesi, the sailfin silversides comprise most of the fish species and probably had few competitors when the first such fish arrived in the lake. *T. prognatha*, mentioned earlier, preys extensively on other fish, unlike the other members of its lineage and most telmatherinids. Hence *T. prognatha* occupies a fairly distinct niche—and has become substantially reproductively isolated from two closely related forms that feed mainly on invertebrates.

The ecological speciation hypothesis lends itself to experimental and comparative tests, and there have now been many such assessments. I became acquainted with one noteworthy study, led by Jelena Rajkov when she was a PhD student in Basel, because I was involved in the editing of her manuscript. Working in and around Lake Tanganyika, she studied the cichlid fish known as Burton's mouthbrooder (*Astatotilapia burtoni*) in river- and lake-dwelling populations as they diverged into potentially distinct species. She tested for lower survival, as predicted by the ecological speciation hypothesis, of river fish relative to lake fish, in lake environments. The cichlids that had been evolving in the lake for generations should have been better adapted to that environment. In addition, she examined the survival of the offspring of matings between these populations; offspring of matings between populations should be ecologically intermediate and survive less well than lake fish, though better than pure (nonhybrid) river fish.

She confirmed that river fish survived poorly in the lake environment, and their offspring, whether pure river or from one river parent and one lake parent, also did relatively badly. When I first read her manuscript, I wondered why there was no reciprocal experiment, for example, with the various types of fish all placed in enclosures in a river. The authors explained the reason in the revised final version of the paper, and I found their comments memorable. They first noted the practical constraints of their remote research site—for example, how hard it was to get to and from. They went on to state that owing to "the presence of crocodiles and hippos in the riverine environment, no reciprocal control experiment in river environment could be performed." Most of my own efforts at field experiments have been with stickleback fish, near Vancouver, Canada. I have occasionally had to deal with a cranky landowner or seagull trying to steal my lunch, but this paper helped me understand one reason there have been relatively few field experiments in the African Great Lakes, especially around rivers. I have seen hippos up close, and they are magnificent but scary. They kill hundreds of people every year, let alone being hard on field enclosures.

One of the most distinctive patterns predicted by the hypothesis of ecological speciation is *parallel speciation*, an idea both clever and audacious. To explain it, let us suppose closely related fish populations are found in physically isolated ponds scattered across a landscape, some with predators, in which a small body size is selected for, and others with no predators and a large body size favored. Parallel speciation occurs if fish from populations with similar patterns of predation and natural selection evolve similar body sizes and other traits, and will mate with one another, whereas populations from different predation regimes will not. In this scenario, not only do particular traits evolve in parallel across parallel environments, so does reproductive isolation. And this is predicted to happen even though all the populations are evolving independently—the bold part of the prediction.

As with ecological speciation, I encountered this idea in an informal presentation by Schluter at the University of British Columbia. Dolph gave a noon-hour talk about this topic, which he had learned of from two studies of fruit flies in which replicated populations evolved for generations in similar or different selective environments, all completely separate with no movement between them. Remarkably, the predicted parallelism in reproductive isolation was indeed observed. When I first heard this idea and these lab findings presented, I was literally speechless. It was just so original and exquisite in how it linked theory and observation. Moreover, it allowed for a definitive test of a truly unique prediction of the ecological speciation hypothesis. This is important because often in science, the most impassable obstacle we face is to come up with a unique prediction of a hypothesis—a testable prediction for which the data can be readily collected. Parallel speciation was all that. Even better, it turned out to be present in nature, not just a curiosity of laboratory fruit flies.

There seem not to have been any full tests of parallel speciation in the major ancient lake systems, but Joana Meier, then at Bern, and her colleagues have proposed a closely related yet decidedly novel scenario for Lake Victoria. They suggest that two pairs of closely related Lake Victoria cichlid species have evolved through *parallel hybrid speciation* (figure 2.2). Using *Pundamilia* species and populations, Meier and colleagues conducted detailed analyses of multiple individual genomes from sister species of *Pundamilia* on several islands, including Makobe Island in the main area of the lake and Python Island, which is in Tanzania's Mwanza Gulf and somewhat isolated from the main body of Lake Victoria.

From their genomic data, they inferred that the *P. pundamilia–nyererei* species pair first evolved in the main area of the lake, including on Makobe Island. The more shallow-dwelling, widely distributed, and likely older species, *P. pundamilia*, which features blue males, then colonized the Mwanza Gulf and Python Island. A little later, individuals from the deeper-dwelling *P. nyererei*, which possesses red-shifted vision and red males, arrived and interbred with the *P. pundamilia*, forming a hybrid population. From this hybrid population, a new shallow-adapted population of *P. pundamilia*–like fish subsequently evolved with blue males alongside a deeper-adapted population of *P. nyererei*–like cichlids with red males. Substantial reproductive isolation also evolved between them.

Because the two species pairs did not evolve entirely independently, they are not quite an example of parallel speciation of the sort that Schluter proposed. And at the genome level, there was a mixture of parallel divergence between the old and young pairs involving the same parts of the genomes, and different regions diverging in the younger pair. But overall, the consistencies across different locations in the lake do suggest an

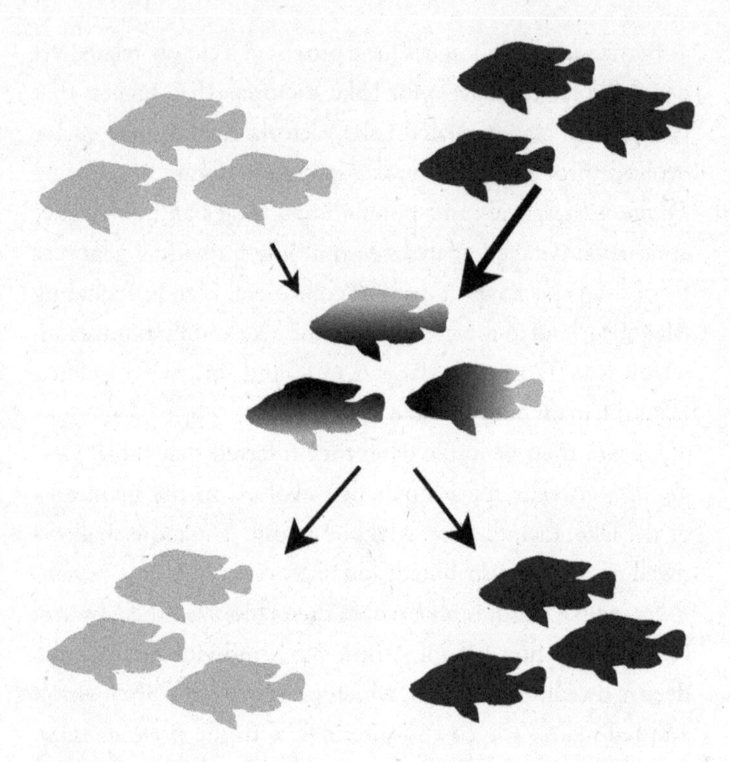

Figure 2.2
Parallel hybrid speciation in *Pundamilia*. *P. pundamilia*, here darker, colonized the Mwanza Gulf and Python Island followed by *P. nyererei*, here lighter. They interbred, forming a hybrid population from which a new shallow-adapted population of *P. pundamilia*–like fish evolved, here darker, alongside a deeper-adapted population of *P. nyererei*–like cichlids, here lighter. *Source:* author, based on figures by Joana Meier.

important role for divergent selection, acting in parallel on similar genetic raw material that became available to the younger pair through hybridization.

ADAPTIVE RADIATION

Ecological speciation is closely linked to the idea and theory of adaptive radiation (though they will not always occur together). Adaptive radiation is an intuitive concept in which a lineage of organisms rapidly diversifies as it adapts to new environments and opportunities. Central to the theory is a guiding role for natural selection in driving divergence in ecologically important traits. It may also contribute to the formation of new species, in particular through ecological speciation. For example, when fish colonize a lake that lacked fish previously, selection often causes some to evolve traits that allow them to feed along the bottom and mainly live in that habitat, while others evolve traits that help them catch tiny free-swimming prey in the open water. Those fish in each habitat that possess genes for traits well suited to their local setting leave more offspring, and genetic differences accumulate over generations—even faster if speciation gets started. Bottom-feeding and open water–feeding fish become more and more different, as each becomes better at feeding and surviving in its preferred habitat.

The term *adaptive radiation* can be thrown around a bit casually, and some use it when discussing any collection of related species. At least in principle, however, a lineage can diversify with a minimal role for natural selection through a haphazard differentiation of physiological, developmental, and other traits that are not always tied to ecology or the environment. This can occur even for animals and plants that live in environments that vary little, especially if they spend their entire lives in a small area, not moving far even when young. If their habitat is patchy, as for some rock-dwelling cichlids in lakes with mainly mud on the bottom and only occasional patches of rock, populations can easily become spatially divided. They can diverge through the processes Mayr emphasized owing to a lack of gene exchange even if selection differs little between, say, rocky outcrops. Thus, sets of related species are not necessarily products of adaptive radiation. Diversification mainly as a result of physical isolation, though, is generally expected to be slower and less predictable.

Ancient lakes are perhaps best known among biologists and certainly among students for providing case studies of adaptive radiation. Biology texts commonly feature a figure showing how cichlid fish in different African Great Lakes have evolved similar forms and appearances as they have adapted to similar diets and ecologies in different lakes. Those of us who have looked at figures of this sort time and again throughout our careers can easily start to take such convergence for granted. But if I stop and reflect just a moment, it really is a wonder. From ancestors as genetically different from each other as we are from apes, fish in lakes hundreds of kilometers apart have evolved strikingly similar heads, body shapes, and fins when confronted by similar ecological challenges. An example of such is shown in figure 2.3.

Some of these patterns are also seen even farther afield. In Sulawesi's Lake Matano, for example, one can find thick-lipped fish from a completely different lineage located thousands of kilometers away, in a lake on a different landmass, which are

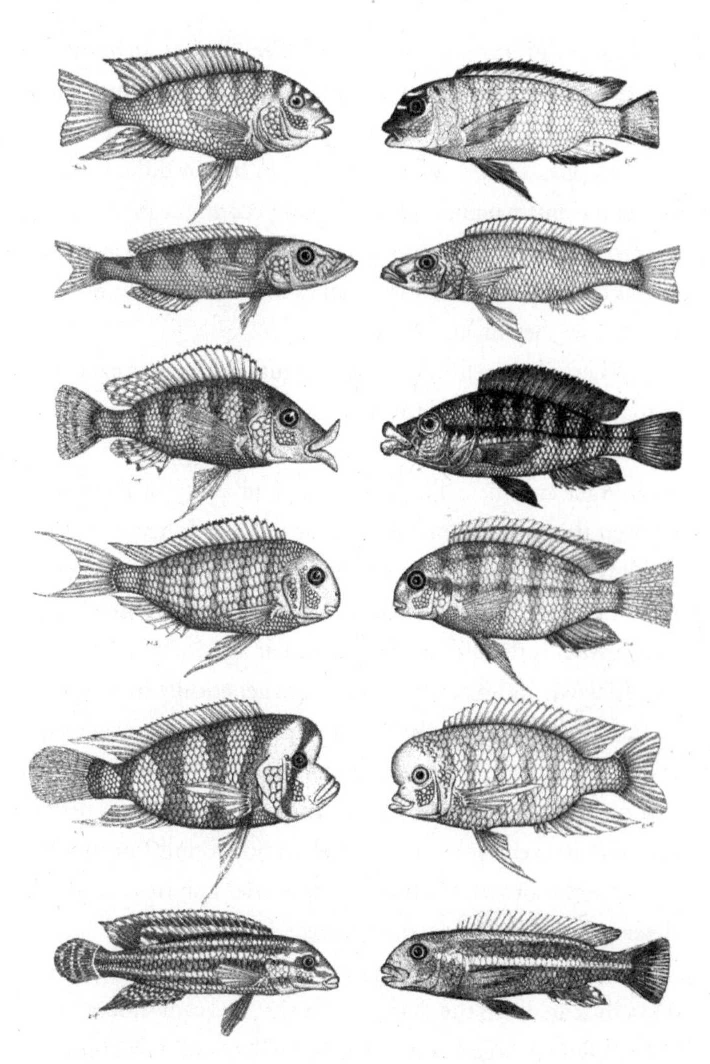

Figure 2.3

Convergent evolution in two ancient lakes. The species in the left column are all from Lake Tanganyika, and the ones on the right are from Lake Malawi. *Source:* Reprinted with permission from Springer Nature, from Albertson and Kocher, "Genetic and Developmental Basis of Cichlid Trophic Diversity," *Heredity* (2006).

reminiscent of forms in the African Great Lakes. In addition, the divergence described earlier, between a shallow-water, bottom-feeding lifestyle versus open water cruising and feeding, is replicated for different groups of fish in different lakes across the entire planet. Just as human economic systems lead to certain similar occupations most everywhere, such as fishing versus farming, fish in different lakes show certain commonalities in how they make a living.

What is less well-known, and arguably even less expected, is that different and relatively unrelated lineages of fish have sometimes diversified in parallel within a single lake. Walter Salzburger and his colleagues at the University of Basel have stressed this pattern for the cichlids of Lake Tanganyika. The cichlids of Tanganyika are special in that they are to the cichlids of the other African Rift Lakes as Africa is to hominins; Tanganyika is the oldest of the trio that also includes Victoria and Malawi, and its cichlids are more genetically diverse than those of the other lakes. In fact, the radiations in the other lakes evolved mainly from lineages within the Tanganyika radiation. The diversity of form present in Tanganyika's cichlids is readily apparent to even the casual snorkeler, and I found Tanganyika's waters and fishes captivating during a brief visit there. It surely deserves more attention from travelers.

Working with Salzburger and Adrian Indermaur, Moritz Muschick analyzed the shapes, ecologies, and evolutionary relationships for a large sample of Tanganyika cichlids encompassing lineages that have been evolving independently for millions of years. Much as for the comparisons between lakes, he documented fish evolving similar feeding habits and adaptations, ranging from jaw shape to body shape, over and over—but these fish, of different species and from different evolutionary backgrounds, were often side by side in exactly the same habitat in the same small area of the lake. This observation raises questions of how they coexist and how they evolved such convergence in the first place. Possibly some convergent forms originated in different parts of the lake, especially during low water periods when Tanganyika was divided into separate basins since periods of temporary fragmentation are not unusual in the larger and older lakes. Ongoing coexistence may have been aided by differences in other key traits; for example, some of the cichlids studied spawn on the lake bottom whereas in others the female carries the fertilized eggs in her mouth. But these explanations do not always apply, and although within-lake convergence is especially obvious and well studied in Tanganyika, it is not unique to this adaptive radiation. Research into this intriguing topic continues.

Although adaptive radiation is a persuasive and intuitively satisfying explanation for many patterns in nature, it is always important in science to conduct decisive tests of our most fondly held explanations. If ecological opportunity mediates diversification, as predicted, it should be possible to identify which aspects of ecology are important and link opportunity to the diversification rate. This has been done in large-scale comparative analyses initiated by Ole Seehausen (figure 2.4), who I have mentioned before. Ole's group has bases both in Bern, Switzerland, and at a research center in the picturesque Swiss village of Kastanienbaum, on the shores of the exquisite Lake Lucerne—which is itself home to radiations of whitefish and charr. Ole is well-known for the intensity of his fascination with cichlid fish, and I had an enjoyable experience of this during a

Figure 2.4
Ole Seehausen in his laboratory in Kastanienbaum, 2019.

visit to Kastanienbaum. It was entertaining, inspirational, and informative to spend a few hours with Ole in his fish lab. He had a marvelous tale to tell about every tank, what had already been learned, and what he wished to do next. He chose his words carefully as he spoke, but also with humor and the joy in the work that is a delight to see in colleagues and students. Later, at a group dinner, I was reminded that even elite scientists are people with families and lives when one of Ole's teenage children joined us. My own eldest had recently moved out of our home, and I felt nostalgic watching the affectionate give-and-take between daughter and father.

Seeking to test hypotheses of adaptive radiation, Catherine Wagner, a young US scientist then working in the Seehausen lab (who later moved to Wyoming), led a series of analyses of cichlid fish and their ecologies across forty-six African lakes. These included the three great lakes we have focused on. Wagner and her collaborators asked what factors made a cichlid newly colonizing a lake more likely to evolve into at least two species, or five or more species. They found that cichlids were more likely to diversify in lakes with characteristics associated with ecological opportunity: greater depth and greater net solar radiation. Depth should provide a wider diversity of habitats to colonize, and more solar energy should support larger populations and/or more species. Surprisingly, lake area was not consistently important in triggering diversification, although large-scale radiations were seen only in the largest lakes. Further, when they looked at the total number of species that had evolved in a lake (a little different from the previous measure), adding up all the different lineages of cichlid present, there were indeed more species in lakes with bigger areas. This could be because of a greater range of ecological opportunities in such lakes, but it could also be because they are large enough for speciation to occur between related populations in different parts of the lake. In lakes in which speciation had taken place, lake age led to greater diversity as well, although the effects of time were not always straightforward.

The Seehausen group has continued these analyses of cichlid diversification in work that expands their data sets to include *every* species of cichlid. This effort was led by Matt McGee, another intrepid US scientist. There are more than 1,700 cichlids (probably quite a lot more; this estimate is likely low), which comprise more than 5 percent of all teleost fishes; teleosts are what we usually think of when we think of fish. Perch, tuna, salmon, and most other fish are teleosts, whereas lampreys, sharks and rays, lungfish, and a few other groups with fewer species are not. For comparison, there are about 6,400 species of mammals on the entire planet. If the 1,700 cichlids had hair and mammary glands, they would represent over one-quarter of mammals. If they had a land-dwelling adult stage, they would account for more than one-fifth of the 8,000 or so amphibians so far described and named (which are mostly frogs, it turns out). This more comprehensive cichlid data set includes fish from South America, India, Madagascar, and a few other spots as well as Africa, and a large number of river-dwelling species in addition to those in lakes. Perhaps it is not surprising, then, that when the relationships between speciation rate and environmental variables are reanalyzed, things have changed a bit.

As far as external factors go, the biggest development in the new data is the appearance of predators. Analysis of this broader data set suggests that when large, visually oriented predatory fish are present, speciation is slower and species accumulate less quickly (note the new analyses are of speciation rate, which is also a little different from examining whether radiations start or not, or looking at the total number of species—though the measures are all related); possibly predators suppress population densities of the species they eat, thereby reducing competition among prey and thus pressure to diversify to obtain enough food. A dry climate also acts as a brake, not surprising for fish. Deep water still facilitates cichlid speciation in the expanded data set, but the effect is weaker. There are additional intriguing correlates, more internal than environmental, which we will explore in later chapters. Nevertheless, environmental correlates of diversification remain in this massive compilation of data in which environmental classifications are of necessity somewhat crude. And these correlates are connected to ecological opportunity. Therefore adaptive radiation is supported, though perhaps less clearly than when one considers only lakes.

Most studies of adaptive radiations and the processes that drive them have either looked at large-scale patterns in traits, ecology, and diversity, or sought to measure selection itself in a much smaller number of species, typically one or two. A study of telmatherinid fish from Sulawesi's Lake Matano took a different and notably ambitious approach in an effort to test adaptive radiation theory more directly and comprehensively. For the five or more incipient species comprising the Matano sharpfin group, Jobst Pfaender and his collaborators collected wide-ranging data from over 1,000 individual fish, including feeding habits, habitat, body shape, and relative liver weight, a measure of body condition thought to be indicative of evolutionary fitness. Evolutionary fitness is essentially the relative contribution to the next generation's gene pool—thus whether a trait is becoming more common or disappearing as a result of differential survival and reproduction. Liver weight is sometimes used as an indicator of fitness because the liver carries energy stores. These stores are an index of the energy available to the fish for growth, reproduction, or other functions, and an indicator of how well the fish is doing day-to-day in terms of feeding success, resisting stress from disease, and so on. It can reveal whether an animal is scrawny and scraping by, or plump and prosperous.

Pfaender and colleagues plotted their liver-based fitness index against measures of shape. These composite shape measures summarized multiple features like relative head width and body depth—features that have often been found to affect how successful fish are when feeding on particular prey types. Such plots are referred to as adaptive or fitness landscapes (figure 2.5). A value for a trait (or set of traits) indicates success on a fitness landscape if it is on a fitness peak, and suggests failure if it is in a valley of depressed fitness. If ecological selection acts strongly in favor of members of incipient species and against hybrids, the average value for each species should be on a peak of fitness whereas hybrid values should have low fitness and be in valleys. This is the pattern seen in Pfaender and colleagues' data, in which peaks are generally associated with sets of features characteristic of members of the five known species while areas between species, where hybrids should usually be located, are generally of lower fitness. Not every peak is exactly where it was expected to be, and some statistical results are not definitive, but the overall pattern is clear and provides satisfying support for hypotheses of ecological speciation and adaptive radiation. The tremendous effort involved in the study turns out to have been well justified.

One of the coauthors on this paper deserves recognition here for the contributions she made to this and many more of the studies on Indonesian lake fishes that will be recounted in this volume as well as the taxonomic work that made some of the theoretically focused studies possible. Renny Hadiaty (figure 2.6) was the systematic ichthyologist at the Indonesian Institute of Sciences Research Center for Biology as well as head of the Ichthyology Laboratory and the curator of the fish

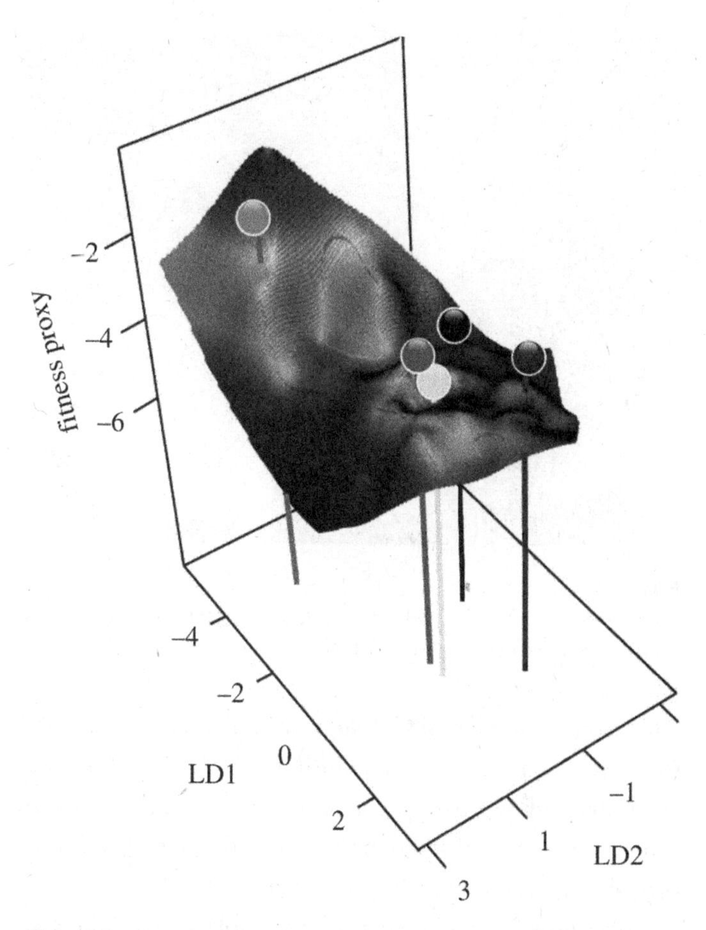

Figure 2.5

Fitness landscape for sharpfin telmatherinid fish from Lake Matano, Sulawesi. The horizontal axes are summary measures of shape, and the vertical axis is a proxy for biological fitness. The "pins" indicate mean values for each species and are mainly on peaks of fitness, with valleys of low fitness between them—where intermediate hybrids generally do poorly. *Source:* Modified from Pfaender et al., "Rugged Adaptive Landscapes Shape a Complex, Sympatric Radiation," *Proceedings of the Royal Society B: Biological Sciences* (2016), under a Creative Commons CC BY 4.0 license (http://creativecommons.org/licenses/by/4.0).

Figure 2.6
Renny Hadiaty (right) with Daniel Lumbantobing and Sopian Sauri during 2010 fieldwork. *Source:* Photo by Lynne R. Parenti, © Smithsonian Institution.

collection at the Museum Zoologicum Bogoriense. In a career that ended too soon with her death in 2019 at age fifty-nine, she published over fifty papers in three languages and made important contributions to the growth of Indonesian fish biology. She also had three fish species named after her!

Our attention is often drawn to isolated radiations as in Lake Matano or spectacular radiations that involve many species as in the African Great Lakes. A quite different pattern of diversification is seen in Japan's somewhat less speciose Lake Biwa, though it occurs to some degree in probably every ancient lake. Biwa is located on the main Japanese island of Honshu, almost midway between the massive urban centers of Nagoya and Osaka, just upstream from the cultural center of

Kyoto. The lake is over four million years old, but underwent substantial geological changes about 400,000 years ago when most of its current deepwater areas were added as a result of a tectonic process known as a *fault-block movement* associated with grabens and half-grabens. These deeper waters, now reaching over 100 meters, are associated with species living entirely in open water, away from the lake bottom. Biwa contains sixteen endemic or semiendemic—confined mainly to Biwa—fish species or subspecies, out of about sixty total species and subspecies (forms that are differentiated, but not quite enough for taxonomists to label them as distinct species).

In contrast to better known ancient lakes, Biwa's endemic species are not mainly the result of a few lineages undergoing repeated speciation events and rapid within-lake diversification. Instead, multiple lineages that are found mainly in nearby rivers have each evolved into one or a few new lake species, often through adaptation to habitats found in the lake yet not present in the surrounding waters. In addition, the lake harbors some lineages now absent from the rest of Japan, or that are genetically distant from other Japanese forms. Thus, Lake Biwa may have served as a reservoir of species and genetic diversity, and made an important contribution to the freshwater fish communities of western Japan outside the lake. This reservoir or refugium (also sometimes described as a museum, in contrast to a cradle) role has also been suggested for some other ancient lakes, at least for some groups of animals. The gastropods (snails) of Lake Tanganyika are a prime example.

Studies of Biwa's snails highlight another important feature of lake radiations: how different lineages can respond differently to the same environments. Snails of the genus *Biwamelania* resulted from two distinct colonizations of the lake. They diversified into fifteen different species as they occupied new habitats and depths following the lake's expansion and deepening about 400,000 years ago. The study of this system is at a somewhat early stage, but it currently looks like spatial isolation may have contributed to speciation in this system in addition to ecological divergence. Some species are confined to individual islets or river mouths, and lack a free-swimming or drifting open water stage that would help the snails disperse. Hence fish and snail diversity have accumulated in different ways in the same lake. We have made important progress in the study of adaptive radiations, but there is much more to come.

* * *

Counting the species in a location is one way to measure biodiversity. Other methods take into account relative abundance, evolutionary history, and ecological distinctiveness. Biodiversity ultimately arises through the evolution of new species, and ecological differences are increasingly considered to be important to speciation, but spatial isolation, time, and chance can play roles too. During adaptive radiations, lineages of organisms rapidly diversify as they adapt through natural selection to new environments and opportunities, and ecological speciation may occur. Ancient lakes provide support for the theory of adaptive radiation and extraordinary case studies. Ancient lakes may also serve as reservoirs of biodiversity, seeding the areas around them.

3 EVOLVING TOGETHER AND APART

The smaller univalve . . . so much resembles a Nerita or Calyptraea that it would be taken for a sea-shell if its history were not well authenticated.
—S. P. Woodward, *Proceedings of the Zoological Society of London* (1859) regarding Tanganyika gastropod genus *Spekia*

When biologists think of adaptive radiation, we generally imagine one species evolving into various new ones in response to ecological opportunity. Opportunity usually means there are few other creatures about and thus plenty of resources, whether because new habitat has opened, such as a lake forming, or competitors have gone extinct. A spectacular example of the latter is found in the opportunities that appeared for mammals when most of the dinosaurs (except birds, which are to dinosaurs as bats are to mammals) were wiped out by an errant asteroid about sixty-six million years ago. But in ancient lakes, other factors can also be important in determining which groups arrive in the lake at all as well as their propensity, once established, to give rise to new species or diversify in form and function. In this chapter, we explore some of these other factors and the intriguing creatures that have resulted. We finish with an account of how even an ancient lake that has vanished may leave a biodiversity legacy.

COEVOLVING SNAILS AND CRABS

While empty habitats can provide opportunities, the presence of other fauna and flora can sometimes accelerate or direct evolution, and may even trigger speciation and diversification. This is especially likely when organisms interact closely, with parasites and their hosts supplying one of the most extreme illustrations. These tight evolutionary interactions, in which species influence each other's evolution, are known as *coevolution*, and can also occur between predators and prey. Hence organisms living closely together and coevolving may evolve in new directions; you might say they evolve apart from their nearest relatives.

In a famous example from Lake Tanganyika, the story begins with some puzzling snails. Gastropods, which are the group to which snails belong, can achieve quite large body sizes in the ocean; conchs and abalone are well-known instances. In Lake Tanganyika, there are freshwater gastropods with thick shells and complex shapes and structures that make them look more like marine species than like the fragile, algae-scraping little animals familiar from many streams, ponds, and lakes. Once Tanganyika's gastropod oddities came to the attention of European scientists, speculations as to their origins quickly began to appear. The first idea was that they were marine relicts left from an ancient connection to the ocean. Another popular notion for a time was that Tanganyika has an unusual, calcium-rich water chemistry that leads to heavier and more ornate shells. Other suggestions were that the spines of some species provide stability on soft bottoms or that thicker shells are beneficial in the shallows because of strong wave action. There were serious problems, however, with all of these hypotheses when they were considered carefully and critically. All were discarded.

A more promising possibility was that Tanganyika snails differed from those in most other freshwater sites because they experienced more ferocious and effective predation, most likely from the heavily clawed crabs known to occur in the lake. In a series of experiments on several species of snails and crabs from Tanganyika as well as localities with more typical freshwater snails and crabs, Kelly West of the University of California at Los Angeles and Andrew "Andy" Cohen of the University of Arizona tested predictions of this hypothesis. They found Tanganyika snails have more "sculpture"—features such as spines and ribbing that they suspected were protective. Tanganyika shells were also much thicker around the shell aperture, the entry and exit through which the snail squeezes much of its body and retreats when threatened. When their strength was tested quantitatively using an instrument from an engineering laboratory, the Tanganyika shells were found to be dramatically stronger than other freshwater shells of similar size. Their strength was more akin to the shells of tropical marine species that are thought to be exceptionally resistant to predation. Also consistent with strong predation pressure, the Tanganyika shells showed much higher frequencies of shell "repair," indicating recovery from damage caused by predators (figure 3.1).

Figure 3.1
Tanganyikan crab *Platytelphusa armata*, showing large, molariform claws, and prey species *Lavigeria paucicostata*, showing unsuccessful crab predation and shell scarring (left shell) and successful, lethal predation (two shells on right). Note shells are enlarged about twofold relative to the crab. *Source:* Reprinted with minor modifications with permission from John Wiley and Sons, from West et al., "Morphology and Behavior of Crabs and Gastropods from Lake Tanganyika, Africa," *Evolution* (1991).

When interactions with crabs were investigated more directly, some of the same features of snail shells were found to enhance survival or force predators into more time-consuming attacks. On the crab's side, bigger claws led to attacks that succeeded more quickly. Moreover, the crabs had evolved claws that were truly impressive—sometimes longer than the width of the crab's shell. These snail-feeding claws had toothlike projections on them that are well suited to crushing. They were broader and sturdier than the "dentition" found on other freshwater crabs, just as our molars are wider than our canines and better at crushing foods such as nuts and seeds. Reflecting this analogy, such claw dentition is known as *molariform*.

When a crab encountered a snail too sturdy for it to crush, the alternative was much less efficient. It would carefully peel the shell back, starting at its opening—a methodical process that required at least forty minutes. Tanganyika crabs are patient!

All in all, it appears likely that Tanganyika crabs and gastropods are so distinctive because of a long history of coevolution in this unusually ancient and stable lake, leading to escalation in predatory traits in the crabs and antipredator traits in the snails. Yet there are only two species of heavily clawed, gastropod-eating crabs in Tanganyika, and some analyses suggest that Tanganyika gastropods, while unusually diverse in terms of their shapes and shell features, have not speciated particularly rapidly. In this case, coevolution may have contributed more to increasing the diversity of forms, and generating unusual ones, at least for a lake, than to quickening the pace of speciation.

A surprisingly similar example of crab-gastropod coevolution has been described from Sulawesi. In each of Sulawesi's ancient lakes, freshwater snails of the genus *Tylomelania* coexist with predatory crabs bearing large claws with, again, molariform dentition. Both are frequently abundant in the lakes. I would see crabs regularly in Lake Matano, and snails were present almost everywhere one looked underwater. The snails are intriguing in several ways—one being their unusual reproduction. They give birth to live young that already possess shells, and their young at birth are the largest known from freshwater gastropods. Their shells, though, show many similarities with those of the shells seen in Tanganyika species preyed on by crabs, as shown by Thomas von Rintelen, Matthias Glaubrecht, and additional collaborators at the Museum of Natural History in Berlin. In each of the snail lineages that have colonized the Malili Lakes, where the most investigation has occurred, lake species have thicker shells than do species found in nearby rivers that lack snail-eating crabs, and shell thickness is correlated with shell strength. Further, about half of specimens from lake species have repair scars, whereas less than 20 percent of individual snails from the rivers possess scars.

The trait most associated with the ecological diversification of the snails, however, is used in feeding. In mollusks, the key feeding structure is the radula, a tonguelike organ covered with rows of tiny teeth, which is used in *Tylomelania* to scrape food from whatever surface the snail is crawling over. Most river-dwelling species have nearly identical radulae covered in small teeth. But in a series of analogous evolutionary events, species that have specialized on hard lake substrates (the bottom material the snails crawl on or through) have convergently evolved strongly enlarged teeth with several distinct forms present.

Species within a lake often differ in the substrate on which they are found, but there is some variation by depth too. In addition to these environmental patterns, spatial isolation has been important to snail diversification given that the species present in nearby lakes are frequently different. Some even have limited geographic distributions within a lake. The genetic basis of radula evolution is now being investigated, and there is some evidence that radula-associated genes have evolved especially rapidly, as expected if they have played a central role in the adaptive radiation of this group. The number of species of *Tylomelania* currently sits at twenty-eight, but some species are highly variable and more may yet be described. Despite the fact that the Malili Lakes are neither exceptionally old as ancient lakes go nor unusually large, this is one of the most extensive gastropod radiations known for a single lineage in a lake system. It seems that the radula of the snail is the beak of the (Darwin's) finch for this inverted archipelago of lakes.

DISPERSAL AND DIVERSIFICATION: SPONGES

One of the most profound decisions an organism makes is choosing where to live. Within our own species, some spend their lives on the block where they grew up, while others leave their hometown, state or province, or maybe even country. I once had a barber, originally from Greece, who had given a good deal of thought to the issue of immigration. Musing on this topic, Andy once told me between scissor clips, "You know, a man who moves to a new country after thirty is no good for either country." I started my shift from Canada to the United States in my twenties, but only finished it in my thirties, so am not sure where I fit in Andy's scheme, but his words often come back to me. Much as for people, how far to disperse is a defining trait for animals and plants, with substantial evolutionary implications. Dispersal usually occurs early in life, but sometimes takes place after maturity.

A long-standing scheme for characterizing the immense diversity of animal dispersal strategies contrasts species that produce armies of cheap, mobile offspring, which are good at finding new homes and opportunities, with those that invest in just a few expensive offspring that do well in dense, competitive environments. Groups with limited dispersal should have lower levels of gene exchange with other populations, and both natural selection and more haphazard local processes should have stronger and quicker effects, leading more readily to the evolution of reproductive barriers and new species. In one truly ancient group of animals—sponges—dispersal capabilities have diverged dramatically among lineages and over time. Some of the most interesting and striking patterns involve ancient lakes.

I first began to give sponges more than passing attention when I visited Lake Baikal. I was on a field trip on the lake with a group of biologists, all of us there for a conference, and we had set anchor in a shallow cove for some diving. One of the divers brought up a bright green, candelabra-like structure, which I initially took to be some sort of plant, but it was in fact an animal—a sponge. I knew sponges could live in fresh water, but the ones I had come across in North American ponds were small, gray, and nondescript. The Baikal sponges had a decidedly marine look to them and were impressive, even more so when I saw photos of them in their underwater habitats (figure 3.2). Since that experience, I have developed a much

Figure 3.2
Sponges of Baikal. *Source:* Olga Kamenskaya.

greater curiosity about sponges and been surprised at what I have learned about their evolution in fresh water. It requires a little general background to explain their presence in Baikal and appreciate their story in ancient lakes.

The sponges are an ancient, mainly marine group and a very early branch off the tree of animals that includes us. Sponges lack organs and have only a few specialized cell types. All sponges have a "skin" of T-shaped or flattened cells and typically an internal system of canals. Lining these canals are specialized cells possessing tiny whips, or flagella, whose regular motions cause water movement. This flow of water allows the sponge to filter out microscopic food particles. Between the skin and flagellated cells is a space containing a protein matrix as well as additional specialized cell types and other microscopic organisms. Despite their structural simplicity, sponges can be meters wide and take diverse forms. They can be crusts on rocks or possess complex structures, sometimes quite rigid and sturdy owing to tiny silica or calcium-based bodies known as *spicules*, and/or proteinaceous fibers of *spongin*. Some sponges are literally rock-hard, with limestone constructions. There are over 9,000 species in total, with many more almost certainly awaiting discovery, as with so many invertebrate groups.

Relative to the vast time frame of their evolution, sponges arrived in fresh water only quite recently, about 300 million years ago, and did not diversify into their modern range and habitats until much later. Whereas many different lineages of fish, mollusks, and other groups have colonized fresh water, with multiple distinct colonizations for each group, just one lineage of sponges has done so. This group, the Spongillida, contains about 240 species, though there are certainly more to be found and named. Members of the Spongillida are not as diverse in form as marine sponges, but still show extensive variation. Yet there is a single trait that seems to have been critical to the success of this lineage. The Spongillida possess a distinctive reproductive body known as a *gemmule*—a mellifluous name that seems to evoke something precious, despite referencing a group of organisms seldom thought of as beautiful or praiseworthy.

Gemmules are small, collagen-covered globules of cells, each with the potential to develop into a mature sponge. They play important roles in both the persistence and dispersal of sponges. Gemmules of some species are tolerant of environmental extremes, including freezing or drying. Some possess chambers that enable them to float and disperse by wind, while others have tiny silica-based spicules that can hook onto flying insects, mammals, and even birds, allowing the minute gemmules to hitch a ride and travel sometimes enormous distances. With such adaptations, freshwater sponges have managed to colonize every continent but Antarctica and evolve into forms ranging from encrustations, which can be challenging to recognize as sponges, to structures more than a meter tall that look like leafless treelets, such as I saw in Baikal. Some individual species are exceptionally widespread, with ranges extending across different continents. Freshwater sponges are even found in desert lakes. Wherever you are as you read this, unless this book has made its way to Antarctica, there could be a gemmule flying over you now, attached to a bird or insect. What a surprising world this is, in which sponges fly around us as we turn pages or tap away at computers.

Gemmules are produced without sex and are in one sense an extension of the ability most sponges possess to grow a new individual from a broken-off fragment. Like marine sponges, freshwater species also reproduce sexually, but the sponge larvae are not good swimmers and typically do not disperse very far.

Although gemmules appear to have played an important role in the success and spread of freshwater sponges and are rare in marine species, they are not present in every member of the Spongillida, the freshwater sponge lineage. The species from which gemmules have been lost, however, seem not to be a random or haphazard assortment of sponges. Gemmules have consistently been lost by species living in particular localities, specifically the ancient lakes, and this loss is evidenced on multiple continents. The lakes whose sponges are known to lack gemmules include Titicaca in South America, Poso on Sulawesi, Ohrid in Albania or North Macedonia, the Caspian Sea in western Asia, Baikal in Russia, and Malawi and

Tanganyika in Africa. Continuing research will surely result in more discoveries of gemmule loss and add to this list.

The geographic distribution of gemmule loss resulted in considerable taxonomic confusion about the relationships among these lake species, at least before DNA sequence data became widely available for more powerful and definitive analyses of relationships. Gemmule-lacking sponges from five of the lakes were initially placed into a single family, the Malawispongiidae, a name referring to the home lake of one member genus. Yet modern taxonomists, the scientists who classify organisms and give them names, almost always insist that classifications be natural. By *natural*, they mean that all species in a group should be descendants of a single ancestor and all descendants of that ancestor should be included in the group, including the ancestor. For the Malawispongiidae, or any classification that put sponges from multiple ancient lakes together by themselves, it always seemed a bit doubtful that these requirements could be met.

In recent years, enough sequence data have accumulated to allow the construction of reliable evolutionary trees for substantial numbers of freshwater sponge species—with complete genomes even starting to appear—and these data confirm the problems with the Malawispongiidae. Rather than most or all the gemmule-lacking ancient lake sponges being descended from a common ancestor that was somehow transported among them, it appears gemmules have been lost time and again from the very freshwater sponges in which they had seemed so essential. Sequence data suggest strongly that the sponges now inhabiting ancient lakes have evolved in a series of independent evolutionary steps from widely distributed gemmule-possessing ancestors. In some cases, they have diversified further within individual lakes, likely subsequent to gemmule loss, with the most noteworthy example being Lake Baikal. In Baikal, fourteen species have been named and assigned to four different genera, all classified within the family Lubomirskiidae. The most abundant species is *Lubomirskia baikalensis*, which forms Baikal's marvelous "sponge forests." In Baikal, these sponges have been estimated to constitute nearly half of the mass of all living organisms inhabiting the lake's bottom. That is an awful lot of sponge, and their volume alone indicates they are important ecological players.

It seems likely that gemmule loss, and the reduction in gene exchange among populations that must result, has played a pivotal role in the evolution of endemic species of sponges in ancient lakes as well as in the further diversification that has occurred in some cases. Unfortunately, the critical analyses have not yet been conducted to definitively confirm or refute such a hypothesis. Nevertheless, the powerful pattern of convergent loss of gemmules in ancient lakes, at distant locations, argues strongly that selection favors this change—that it is an adaptation. The presumed explanation for repeated gemmule loss is that the stable, predictable environments provided by ancient lakes reduce the benefits of producing a structure specialized for wide dispersal and survival through severe disruptions such as the habitat drying out. Evolutionists think of most every trait as having a cost of some sort, and that will certainly be true for a form of reproduction like gemmules, so when it offers little benefit, gemmule production should be selected against and eventually be lost. The strong convergence observed here also argues for important environmental similarities among even distant ancient lakes, at least from the perspectives of the organisms inhabiting them. To a sponge, it seems to be critically important whether a lake is of the ordinary sort or ancient.

Additional data will be needed to test hypotheses about the evolutionary advantages and disadvantages of convergent gemmule loss. Basic taxonomic work is needed too, but this may not move quickly because freshwater sponges are a neglected group, with few scientists specializing in them—quite possibly fewer, for example, than work on the single species of Atlantic salmon. Things may be looking up, though. Sponges have been attracting more interest from bioprospectors, who assay extracts from sponge tissues in search of new compounds with potential applications in the treatment of disease or use in biotechnology. Many of the compounds of interest are not produced by the sponges themselves but rather by bacteria and other organisms that live within the sponges, and may have been coevolving with sponges for a long time.

SHRIMP RESPLENDENT AND RADIATING

In the freshwater shrimp of the family Atyidae, found in several ancient lakes, diversification has been proposed to result from an interaction between opportunity and dispersal more subtle than is seen in the sponges. This widespread, almost exclusively freshwater group has undergone radiations in Sulawesi's lakes and Lake Tanganyika, with the Sulawesi forms more extensively studied. Most of the work has been led by Kristina von Rintelen at the Museum of Natural History in Berlin. There have been two von Rintelens working on Sulawesi invertebrates; the work on snails discussed earlier in this chapter was done mainly by her husband, Thomas. Their fieldwork has sometimes been a family project, with their two daughters helping out on one trip. For their contributions, and to highlight the importance of the shrimp for coming generations, each daughter had one Lake Poso shrimp named eponymously for her: *Caridina lilianae* and *Caridina marlenae*.

Atyid shrimp have diversified in both Lake Poso and the Malili Lakes (figure 3.3), but their investigation has been more extensive in the Malili Lakes. Fifteen endemic species are found in the Malili system, with thirteen in Lake Towuti. I would see them sometimes when I worked in Lake Matano, peering out from between the rocks as they poked daintily at the algal communities growing around them. They looked impossibly fragile, and I marveled at their ability to persist in the presence of larger crabs and fish, which seemed loutish by comparison with the elegant atyids.

Some species (most, really) are arrestingly attractive with complex, unexpected patterns of stripes, bands, and spots of most any hue one might imagine, and differing greatly from one species to the next. *Caridina profundicola*, for example, has a subtly yellow background color with tasteful, slim orange stripes stretched side to side here and there across its body. In some females, this pattern is paired with eggs of a blue-green hue that seems unnatural, as if photoshopped to contrast with the mother's body. The aptly named *Caridina striata* possesses a series of alternating red and white stripes running from its head to the end of its tail. Although eye-catching in a photograph, von Rintelen and Yixiong Cai (in a 2009 paper naming several new species and generally reorganizing the group; Cai is based at Singapore's National Biodiversity Centre) comment that the

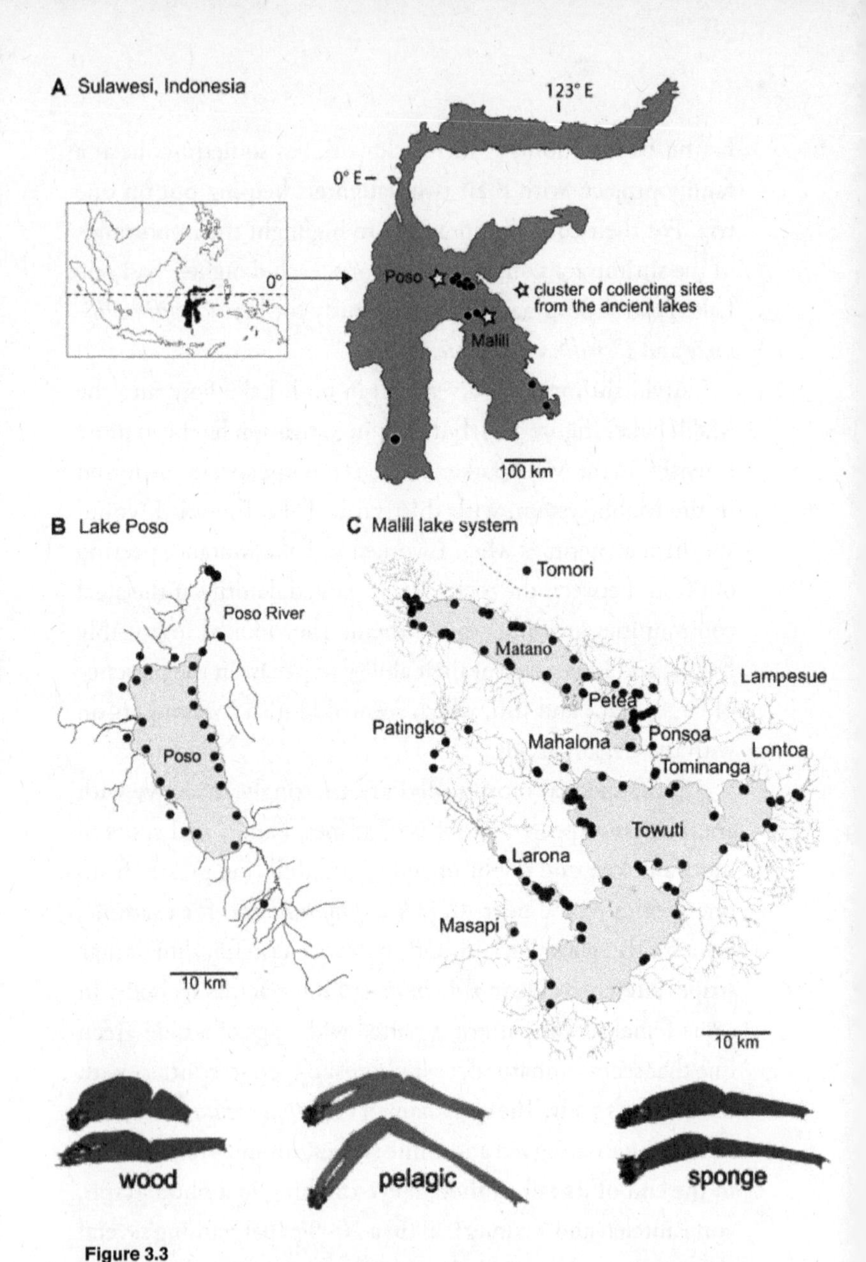

Figure 3.3

Map of Sulawesi ancient lakes with shrimp collecting sites indicated (*left*) and examples (*right*) of shrimp chelipeds of species found in different habitats. *Source:* Reprinted with minor modifications with permission from John Wiley and Sons, from von Rintelen et al., "Adaptive Radiation and Ecological Diversification of Sulawesi's Ancient Lake Shrimps," *Evolution* (2010).

body of this species, as for some others, can be inconspicuous in natural settings. Its white chelipeds, however, the long set of legs ending in pincerlike claws, are always clearly visible. While reds and oranges are prominent in the color palette of the Malili Lake shrimp, blue is rarely seen on their bodies. Thus, the azure patches on Lake Poso's *C. caerulea* are so noteworthy as to inspire its name—*caeruleus* means blue in Latin. The blue appears in two almost-glowing ovals on the final segments of its tail, part of a fringe of blue comprising also its legs and antennae, all providing a vivid frame for a peach body. Most species, regrettably, are listed as critically endangered on the International Union for Conservation of Nature Red List.

Von Rintelen has investigated the anatomies and distributions of these shrimp in detail, and conducted analyses of their evolutionary relationships as well. In contrast to early suggestions that the anatomy of shrimp was unrelated to their feeding habits and ecologies, von Rintelen observed strong relationships. The first and second pairs of chelipeds possess paintbrush-like structures that are used, in most species, to brush fine food particles from the substrate. Just as teeth differ among mammals depending on whether they use them to grind plant matter or cut through animal flesh, chelipeds vary among shrimp living in different habitats and eating different foods. In the case of the Malili Lakes shrimp, the differences among species with different ecologies are obvious, and much clearer to even a casual glance than the shape differences among members of some fish radiations. The chelipeds of open water (pelagic) species, which are elongate and delicate, are especially distinct from the short, burly chelipeds of species found on wood substrates (figure 3.3).

Not all shrimp species were confined to a specific habitat, however. Generalist and river-dwelling species could occupy multiple ecological settings. Von Rintelen and colleagues suggest that differences in habitat flexibility have affected the propensity to disperse; a species with narrow habitat needs will more often encounter unsuitable areas across which it cannot disperse, in particular the rivers that link the lakes. Such strong habitat preferences may, as a result, cause one species to perceive a barrier where another does not, and lead to speciation through geographic isolation where a barrier is perceived. Similar suggestions have been made for other groups in other lakes. In the African Great Lakes, for example, some of the cichlids have inflexible habitat requirements, especially those species confined to rocky outcroppings that are separated by sand or mud. The sand or mud may be an insurmountable barrier to the rock dwellers, and populations separated by an estuary or small bay may exhibit extensive genetic differentiation as well as color pattern differences. Yet fluctuations in lake levels, which appear to have been over 400 meters in as little as the last 100,000 years, provide a complication. When such massive changes occur, the formerly separated rock-dwelling populations may come into contact, putting a stop, even if temporary, to further differentiation. Conversely, lake basins that are connected when water is high can be separated from each other, promoting divergence between sets of populations isolated by that process.

At least one shrimp has a particularly unusual habitat requirement for a freshwater species, and a notably constrained geographic distribution as well: *Caridina spongicola* lives exclusively in and on a freshwater sponge in Lake Towuti. Both sponge and shrimp are found only in a single bay, which leads

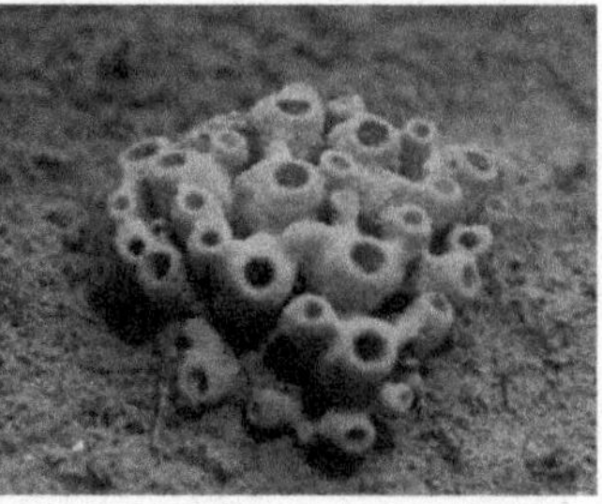

Figure 3.4
Atyid shrimp, *Caridina spongicola*, and the sponge with which it is commonly associated in Lake Towuti (note shrimp at greater magnification). *Source:* Used with minor modifications with permission of the Royal Society, from von Rintelen et al., "Freshwater Shrimp–Sponge Association from an Ancient Lake," *Biology Letters* (2007); permission conveyed through Copyright Clearance Center, Inc.

to Towuti's outlet river, at depths of about three to ten meters. The sponge is so little studied that it does not yet have a Latin name, but based on DNA sequence data and its visible features it is clearly distinct from Sulawesi's other freshwater sponges. It is a respectable size for a freshwater species, up to about twenty centimeters across, and possesses a cluster of distinct cavities similar to those of many marine sponges (figure 3.4). The cavities are where the shrimp are mainly found, with as many as 137 individuals on a single sponge. The shrimp don't seem to be eating the sponge, nor do they appear to be providing it any benefit. Probably the relationship is a "commensal" one, in which the shrimp gain shelter and perhaps a food supply inside the sponge, without causing the sponge any substantial harm or paying it any sort of rent. They are a bit like well-behaved squatters in an empty building, although these squatters seem to be feeding on microscopic algae, which accumulate in the cavities of the sponges.

In marine systems, shrimp live in or on a remarkably long and diverse list of creatures, with which they seem in many cases to be coevolving. These range from corals to sea urchins to oysters to sea cucumbers as well as sponges, of course. Such relationships are rare in fresh water, which prompts the question of how a shrimp came to be living in a sponge in Lake Towuti. Von Rintelen and her colleagues have proposed a hypothesis for the origins of sponge dwelling, again connected to the dynamic water levels of lakes, and based on their observations of the shrimp and sponges over several years and a wide range of climatic conditions. Although commonplace lake-level changes of one to two meters seemed to have little effect on the shrimp or their habitats, in a year with exceptionally low water levels, some shallow rocks were exposed to the air. These typically provide important habitat for rock-dwelling shrimp species, which had to find new homes. That same year, rock-dwelling shrimp were observed alongside *C. spongicola* for the first time on sponges that were not as far below the surface as they had been. Von Rintelen and colleagues suggest that such a shift may have contributed to sponge dwelling, and even now may facilitate hybridization between shrimp species normally found in distinct habitats.

The presence of freshwater shrimp in an apparently obligatory relationship with sponges is also noteworthy for the intriguing possibilities it raises. In the sea, sponge-dwelling shrimp have repeatedly evolved a reproductive and social system that is more familiar from ants, bees, and termites, known technically as eusociality. *Eusociality* is usually defined as involving the cooperative care of offspring, including the offspring of other individuals (common in our own species, with day care an obvious modern example), overlapping generations, and a division of labor between reproductive and nonreproductive individuals—queens versus sterile workers in honeybees, for instance. Eusociality has been suggested to be a key to the enormous ecological success of ants, bees, and termites, but until it was discovered in the marine snapping shrimp genus *Synalpheus*, it was known only from land-dwelling groups. There are as yet no published reports on the social system of *C. spongicola*, and quite possibly it is entirely ordinary . . . but maybe those 137 sponge-cohabiting shrimp are doing more than just sharing a squat.

CAN AN ANCIENT LAKE EXPLAIN FISH DIVERSITY IN SOUTHERN AFRICAN RIVERS?

So far in this chapter we have been seeking to understand how coevolution with other species, dispersal, and ecological opportunity each may have contributed to biodiversity in ancient lakes, and how they can interact. We finish by reversing that approach and ask if an ancient lake might leave a legacy of biodiversity even after its depths have become dry land. If true, then the diversity that evolves in one radiation may enable a wider range of forms in a new setting.

This tale begins with Hull University's Domino Joyce, who together with an international group of collaborators sought to explain oddly high diversity in the cichlid fish of southern African rivers, relative to rivers farther north. Such a pattern is surprising because of the well-documented tendency for biological diversity to increase as one approaches the equator and diminish as one approaches a pole. Yet in the rivers abutting Lakes

Malawi and Victoria, the diversity of haplochromine cichlids, which radiated so explosively in those lakes, is low—typically just one species from each of the two major haplochromine lineages. In contrast, river systems a little to the south (farther from the equator) as well as two to the west usually contain several haplochromines and in one case at least twelve. Beyond their impressive species numbers, the river-dwelling cichlids of southern Africa occupy an unexpectedly broad range of ecological niches, with a correspondingly enhanced diversity of body shapes and feeding traits. At the same time, their DNA sequences suggest evolutionary relationships among these fishes that are not easily explained.

By combining analyses of evolutionary relationships of the riverine cichlids with data on their distributions, Joyce and her colleagues arrived at the surprising conclusion that their exceptional diversity is a persistent signature of evolutionary events that took place in an extinct lake, Lake Palaeo-Makgadikgadi. Until about 2,000 years ago, a lake at times larger than Switzerland occupied a portion of Botswana where there is now mostly dry salt pan or seasonal lagoons. Like Victoria or Malawi, this immense body of water provided a wide range of ecological opportunities during its periods of greatest size and depth. The "megalake" periods were somewhat intermittent, though. Interspersed with periods when the lake may have fragmented and even dried completely, their transient nature presents a potential issue with this hypothesis. Nevertheless, when Joyce and colleagues mapped the distributions of the region's river-dwelling haplochromine cichlids, their highest diversity centered on the location of the former lake. This was not true for other fish groups, in particular members of the genus *Barbus* (familiar to aquarists as *barbs*), which is the most species rich of the mainly river-dwelling African fish groups. *Barbus* showed a more familiar pattern of increasing diversity toward the equator, with a jump in numbers in the Congo River, and no peak at the location of Lake Palaeo-Makgadikgadi.

The range of body shapes present in the river radiations of haplochromine cichlids were revealed, by a quantitative analysis, to be similar in scale to what is seen in Lakes Malawi and Victoria, and much greater than what is present in the river species of East Africa. Certain combinations of traits were missing, particularly those characteristic of fish that feed in the open waters of lakes. The authors suggest that these may have evolved in the extinct lake, but that fish possessing such open water–adapted traits were unable to transition to rivers that lacked appropriate habitat—whereas snail eaters, for example, could find prey in a river just as they had in a lake.

Joyce and her colleagues conclude that the ecological opportunities afforded by Lake Palaeo-Makgadikgadi enabled the generation of ecological and anatomical diversity in the lake's haplochromine cichlids of a sort that would not have appeared in rivers—but that large rivers provide sufficient ecological variety to maintain a great deal of this diversity once it has evolved. African riverine cichlids are currently being investigated extensively using larger and more robust DNA data sets because a better understanding of the evolution of river species is widely perceived as important for interpreting cichlid evolution in Africa's lakes. It will be fascinating to see if Joyce and her colleagues' hypothesis holds up, and if indeed an ancient lake may sometimes generate a multiplicity of forms and ecologies that persist even after the lake's waters have long since departed.

* * *

Ecological opportunity is central to adaptive radiation, but which groups arrive at an open habitat, which other groups they coevolve with, and additional factors may also influence the rate and direction of evolution. In Lake Tanganyika and Sulawesi, predatory crabs and their snail prey have coevolved and converged. In ancient lakes from Africa to South America, stable habitats have led freshwater sponges to lose their dispersive gemmules, facilitating divergence. Atyid shrimp in Sulawesi have similarly diverged in dispersal tendencies as they have adapted to different lake habitats. Puzzling fish diversity and distributions in southern African rivers did not result from the extraordinary dispersal abilities of the groups involved but instead may have arisen through "seeding," by a now-defunct ancient lake and its adaptive radiations.

4 SPECIES IN THE EYE OF THE BEHOLDER

Lake, oh fresh lake of rivulets in-land sea . . .
In riches is Africa, Malawi's reigns of all lakes
—Paul Mwenelupembe, 2013

If your memory stretches back to incandescent bulbs, you will know how different a room looks and feels with the warm, somewhat-red tones of incandescent lighting relative to more sterile "cool white" fluorescent lights. You may also have noticed that the colors of paint and furniture can seem different depending on the light source. Our brains compensate somewhat for this, but these effects are still detectable and at times quite strong. This is because the color of an object is a result of both the wavelengths of light hitting it and which wavelengths are most likely to bounce off it (and in which directions, though we will not explore that point for now) versus being absorbed. Thus, in terms of the environment's role, an animal's color is quite different from the sounds it makes. We produce sounds, but only reflect the light that conveys color. It is a bit like the difference between a motorboat, which propels itself, and a sailboat, which has to work with the wind.

As with artificial lighting, environmental light in nature varies with the light source as well as the weather and time of day. Habitat too has an effect; forests, for example, have green-tinged light in their interiors. We perceive such differences because our eyes contain cones that are sensitive to long (red cone), intermediate (green cone), or short (blue cone) wavelengths of light. When one or two cones are stimulated more and the other(s) less, we perceive color rather than white, black, or shades of gray. But not everyone's perception is the same. For instance, those with red-green color blindness are typically missing either a functional red or green cone and can completely miss colors and patterns that are obvious to everyone else—something it is wise to take into account when choosing colors for a graphic or illustration. Much variation in color vision is based in genes that encode the different opsin proteins that are characteristic of each cone type.

Underwater, ambient lighting differs more from place to place than it does in terrestrial environments. This is because water generally absorbs much more light than does air, and different wavelengths are often absorbed unequally, affecting the light spectrum. These effects can vary depending on how much sediment and other materials are in the water.

Ole Seehausen recognized early on that variation in underwater light environments could have major implications for speciation and species persistence in groups for which color is important in communication and reproduction. In an influential 1997 study, he and two colleagues presented a novel data set that provided evidence of such effects. Their focus was three groups of rock-dwelling cichlid fish from Lake Victoria. Males of closely related species in each group typically have different coloration, with one species mainly blue, and the other mainly red or yellow; these colors are a good match to the peak sensitivities of cones in the eyes of the fish. Distinct color patterns are also present within some species, especially for males, once again usually blue individuals versus others with differing amounts of red or yellow. Conveniently for Seehausen's study, these fish inhabit rocky bottoms associated with islands that are frequently separated from each other by long stretches of sand or mud that are inhospitable to the fish. Thus, the populations and fish communities on each rocky area are substantially independent, making them well suited to analyses comparing the characteristics of different populations.

For fish from thirteen rocky islands in southern Lake Victoria, Seehausen and his colleagues documented an unexpectedly strong relationship between male color, the number of species present from these three groups of cichlids, and the turbidity of the water. Where the water was more turbid, or cloudy, and the light hitting the fish comprised a reduced range of wavelengths, there were fewer distinct male color patterns present and males more often exhibited duller hues. It seems there is little point having intense blue color—reflecting light from the short wavelength end of the spectrum, the blue portion—if there is no such light in the environment. The same would go for the long wavelength red light at the opposite end of the spectrum. Further, there were fewer species of each group at the islands with a narrower range of light (figure 4.1).

Seehausen and his colleagues suggest that the reduction in the number of species at narrowly illuminated sites results from a breakdown in mating isolation. In a light environment that reduces the ability of females to perceive color differences

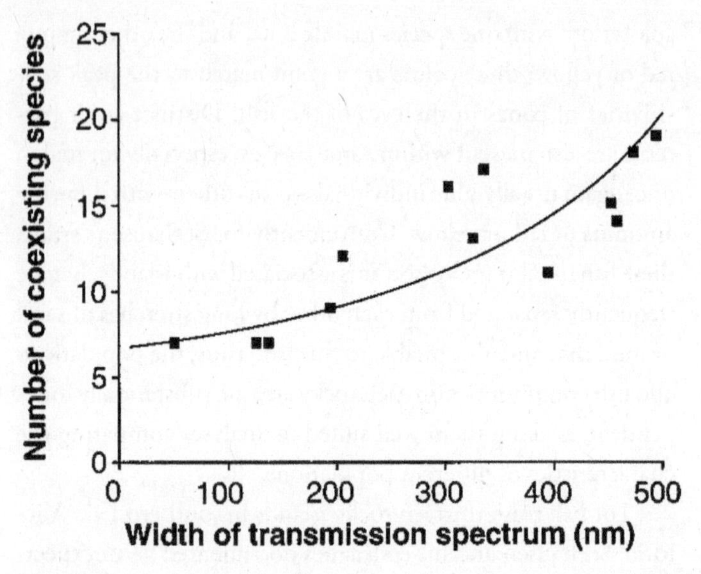

Figure 4.1

Relationship between the width of the underwater light spectrum at Lake Victoria rocky islands and the number of species present. *Source:* Reprinted with minor modifications with permission from AAAS, from Seehausen et al., "Cichlid Fish Diversity Threatened by Eutrophication That Curbs Sexual Selection," *Science* (1997).

between species, it becomes harder to identify their own males—a difficulty that increases further as male color patterns evolve to be less distinct. Mismatings result, and because intrinsic incompatibilities are largely absent in this young radiation, two well-differentiated species can quickly become a single, somewhat variable one, or one species can be essentially absorbed into another. Seehausen's group has obtained additional support for this interpretation in laboratory experiments in which the light spectrum is manipulated.

This is a significant conservation issue because human activity causes much of the turbidity that affects what fish can see. Historical data indicate that the cutting down of forests and changes in farming practices have been causing Victoria's water to become murkier since at least the 1920s. When we think of extinctions caused by human activity, we usually think of overexploitation, like what happened to the passenger pigeon, or catastrophic consumption by introduced predators, such as the effect of feral cats and foxes on Australian marsupials. Habitat destruction also comes quickly to mind, as for the enormous numbers of tropical forest species now extinct, or nearly so, following the conversion of forests to palm plantations and pastures. But with turbidity, human-initiated environmental changes often lead different species to merge rather than one or both disappearing in the familiar way—yet biodiversity is just as surely lost as through better known processes. These effects are not confined to fish or even vision. Whether through increased underwater turbidity, elevated noise in cities, or explosions in the deep sea from weapons tests, disruptions of animals' sensory environments by human activities are increasingly acknowledged to be a significant conservation problem.

COMPETING FOR MATES: SEXUAL SELECTION

The findings from Lake Victoria highlight the importance of competition for mates in the evolution and persistence of species, and emphasize how the environment can mediate such competition. Selection of this sort, known as sexual selection, is the solution that Darwin identified to a difficult challenge to his theory of evolution by natural selection: the puzzling existence of conspicuous traits—like the absurdly large tails of peacocks or vibrant colors of some fish—that seem unlikely to help animals survive better or be better parents. Frequently the opposite appears more likely: a brilliantly colored tail longer than one's body should attract the attention of predators and make it harder to escape from them—and such costs have now been documented in studies with a wide range of animals. In addition, extravagant, expensive traits may disadvantage an animal trying to catch prey or provide for its offspring. All else being equal, conspicuous and costly traits would seem destined for quick elimination by natural selection, which made their ubiquity in nature a long-standing puzzle.

Darwin hypothesized that although potentially costly in terms of survival, ornaments could be advantageous in the pursuit of matings. So even if a long tail makes predation more likely, if females reject males lacking these cumbersome appendages, the males contributing their genes to the next generation will be the ones with long tails—and their sons will also have such tails. Thus, sexual selection, in which males with conspicuous, exaggerated traits get more matings, could compensate for natural selection against such ornaments and lead to their evolution.

Comparative analyses suggest that sexual selection has contributed, alongside divergent natural selection, to adaptive radiations of cichlids. Catherine Wagner and colleagues' analyses, reviewed in chapter 2, also reveal that color differences between the sexes are positively associated with diversification. These differences function as a proxy for the strength of sexual selection in some comparative studies and are thought to indicate strong sexual selection on males. Patterns suggestive of sexual selection's influence are weaker in Matt McGee and colleagues' larger survey.

There is also a good deal of evidence from African Great Lakes' cichlids that not only does the overall hue of a male's coloration—that is, whether it is red or blue—influence his success at obtaining matings but the intensity of his color is important too, at least in suitable lighting. Mike Pauers, then at the Milwaukee campus of the University of Wisconsin, conducted one of the first studies documenting such effects. I helped Mike with some of this work and got to know him along the way. He studied the Lake Malawi species *Labeotropheus fuelleborni* (since renamed *Labeotropheus chlorosiglos*), which has a complex color pattern including an orange belly and sky-blue back. Mike showed that females prefer males whose orange and blue patches are intensely colored rather than washed out. His results also suggest that females prefer males with greater dissimilarity between their various color patches—as if the more striking contrast between a vivid orange and intense blue, compared to the contrast between a dull orange-brown and faded blue, was attractive in and of itself. It is reminiscent of how people sometimes choose articles of clothing, selecting a skirt or tie to contrast strongly with the blouse or jacket next to it, for a more eye-catching overall effect.

Pauers's fascination with sensory biology and the aesthetics of animal social preferences likely has roots in his second career, based in nightclubs and recording studios rather than laboratories and classrooms. Mike is an accomplished saxophone player, recording and performing with a jazz ensemble as well as joining in with touring acts in need of a skilled saxophonist. This is not as unusual as it might seem, as I have often seen artistic and scientific creativity side by side in colleagues, and their families. There seems to be great overlap in the wellsprings of creativity, regardless of the form in which it is expressed.

I have so far focused on sexual selection on males, and that is the norm in evolutionary studies because males more often display extravagant ornaments and perform strangely complex, conspicuous displays. The most widely accepted hypothesis to explain this pattern is based on investment: simply put, sperm are small and cheap whereas eggs are big and expensive. Consequently, in the marketplace of mating, there usually should be an excess of sperm available and males ready to mate, while eggs will take longer and more resources to generate. Other factors also influence the costs and benefits of mating for each sex along with the direction of sexual selection, but investment is central.

Since sexual selection concerns mating success and patterns, it makes good sense that it could play a role in speciation. It was not until late in the twentieth century that this possibility started to attract sustained attention from researchers, but the scientific community had been aware of the issue for considerably longer. One of the most important ideas about how sexual selection could accelerate the evolution of reproductive barriers came from the famous statistician and mathematical biologist R. A. Fisher. In the first decades of the twentieth century, Fisher hypothesized a mechanism by which the rate of evolution by sexual selection could be rapid indeed and proceed to almost shocking extremes.

Fisher, however, was more than a talented theoretician. He was also a eugenicist, and a case study in how science in the twenty-first century is struggling with the challenge of giving appropriate credit for scientific contributions while rejecting destructive or unethical ideas and behaviors. Eugenic thinking has led to forced sterilization and worse, and because of Fisher's support for eugenics, his name is being removed from awards, buildings, and so on. It is a difficult legacy, but one which is crucial to acknowledge.

THE RUNAWAY HYPOTHESIS OF SEXUAL SELECTION

Fisher pointed out that the strength or even presence of a female preference for, say, males with long tails will generally vary among individuals, and such variation will usually have a genetic basis. Some females might carry an allele that causes them to mate only with long-tailed males, while others might have alleles and preferences for short-tailed males or be ambivalent about tail length. Tail length, or whatever conspicuous ornament is the target of the preference, should be genetically variable as well. With both traits genetically variable, it follows that when a long-tail-preferring female produces offspring, they will possess alleles for both the preference, inherited from their choosy mother, and long tails, inherited from the long-tailed male their mother chose to mate with, their father. Daughters of this union will choose long-tailed partners, as their mother did, and sons will express their dad's long tail (though each sex will carry alleles for both traits). As time goes on, a population-wide pattern should emerge for individuals with alleles for long tails in males to also carry alleles causing a *preference* for such tails.

Where this gets really interesting is if long-tailed males are favored in the population overall, which could cause the pace of change to accelerate. This might occur if the preference for long tails is a little more common than that for short tails, or maybe long-tailed males are carrying other beneficial alleles, sometimes known as "good genes." In any case, if the long tail allele is favored, it will get a little more common each generation—and since males with long tails usually carry an allele for the long tail-preference, the frequency of the preference will get pulled upward. With the preference getting more common, the benefit of having a long tail keeps increasing too! Thus, long-tail success begets a stronger preference begets still more long tails, and so on. This self-reinforcing process is sometimes described as *runaway* sexual selection. It can result in not only rapid but also somewhat arbitrary changes in both male ornaments and female preferences.

The female preferences and male traits that evolve to extremes through the runaway process may not be the same in different populations. Chance typically plays some role in evolution, whether through which mutation happens to appear first or other random processes that become especially significant when populations are small—and the runaway process has immense potential to amplify such random differences. The important insight here is that if the potential for sexual selection is high, even briefly separated populations can quickly evolve different preferences and ornaments and come to reject each other as mating partners. Some modeling suggests that such processes can even propel a single population to divide into two daughter species without any physical barriers.

With the development of some of the first thorough mathematical models of runaway sexual selection in the 1980s, enthusiasm was so great that some suggested the runaway process should be the default explanation (the *null hypothesis* is the five-dollar term scientists usually use) for extravagant male courtship traits and female preferences for them. Speciation by runaway sexual selection was also championed as a possible explanation for the mind-boggling rates of speciation that were starting to be confirmed through molecular studies of some ancient lake radiations. Genetic patterns consistent with the runaway process, though possibly not unique to it, have indeed been observed in natural populations including freshwater fish and in laboratory evolution studies. But further models have called into question whether this process will play as big a role in divergence as was supposed, and almost forty years after runaway models first began to appear, there are no examples of speciation in nature that we can point to as clearly the result of such sexual selection. Ruling out competing explanations for the evolution of female preferences is always difficult, so it is important to acknowledge that the absence of evidence in this case may not be definitive evidence of absence. In addition, runaway-like processes may complement and augment ecological divergence, thereby contributing to speciation without being the sole driver; indeed, sexual selection is generally thought to be most likely to contribute to speciation when acting in concert with natural selection. A different approach to the evolution of mating preferences and male display traits has, however, lately supplanted runaway as a research focus, especially in the context of speciation.

SENSORY DRIVE

As we have seen, Seehausen and his colleagues provided persuasive evidence, starting in the late 1990s, that water turbidity can influence cichlid mating patterns and the expression of mating preferences. When water clarity deteriorates enough, speciation can even reverse itself, and two well-defined species can collapse into a swarm of ill-defined hybrids. But there is more going on. In the 1980s and 1990s, shortly after the study of sexual selection began its resurgence, the role of sensory systems in sexual selection started to receive more attention. One of the leaders in this body of work was (and remains) John Endler, a scientist who hails originally from Canada and has worked extensively on guppies and bower birds, among other groups. I first met John while a graduate student, and was impressed (as I still am) with the breadth of his knowledge and interests as well as how he was able to link the details of natural history and physiology with mathematical models of genetic processes, generating unexpected insights along the way. When John was at the University of California at Santa Barbara, I once spent a happy Saturday afternoon with him while visiting to get help with an instrument calibration. We drank tea and watched hummingbirds court from the balcony of his house, which was set on a hillside and surrounded by trees. His enthusiasm for closely observing the behavior of these birds was contagious, and I have no doubt that such habits of careful observation in natural settings helped lead him to the highly synthetic sensory drive hypothesis that emerged from work by himself and several other scientists.

Sensory drive is, in essence, the hypothesis that environmental factors play a central role in the evolution of both mating preferences and ornaments. In the case of mating preferences, this will typically occur when sensory systems are molded by natural selection to function effectively in an organism's particular sensory environment. The strongest selection is expected to be for finding food and other activities directly related to survival, with effects on mate preferences often a by-product. Ornaments that have evolved to attract mates should be influenced by both the sensory biases of receivers and the direct effects of the environment on which types of ornament are most conspicuous. In the case of visual signals, these effects can arise through the spectra of light that illuminate ornaments as well as through the backgrounds against which courting males are seen and extent to which ornaments contrast with those backgrounds (figure 4.2).

Working in Sulawesi's Lake Matano, Suzanne Gray, along with several collaborators including myself, found evidence for the importance of contrast with the background in a comparison of different populations of a fish, *Telmatherina sarasinorum*, in which five different male color patterns are present. She first evaluated *Telmatherina* vision using a closely related species and characterizing the types of cones present in their eyes and their sensitivities. She then used those data to evaluate the conspicuousness of each type of male in each of the two environments in which they courted and spawned, using careful measurements of the light reflected from the bodies of the males in addition to measuring the background light spectrum. She included ultraviolet light as well as wavelengths visible to our species, since these fish, like many animals, can see in ultraviolet light. In one environment, spawning took place on roots hanging in relatively open water with a mainly bluish

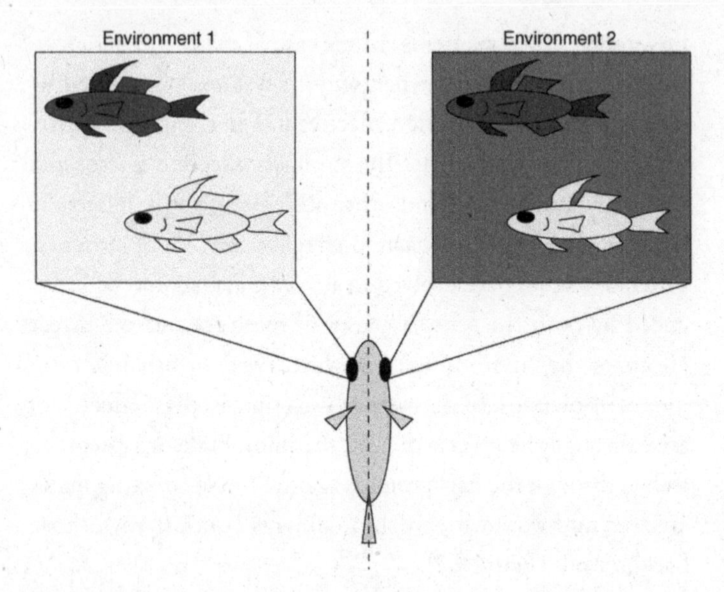

Figure 4.2
Conspicuousness can vary with color and background. Here lightness versus darkness is illustrated, but examples discussed in this chapter mainly concern hue, such as for blue- versus yellow-colored fish. *Source:* Reprinted with minor modifications with permission from Elsevier, from Gray and McKinnon, "Linking Color Polymorphism Maintenance and Speciation," *Trends in Ecology and Evolution* (2007).

background. In the other, fish spawned in shallow water where they saw each other mainly against the typically yellow-brown rocks. Blue and yellow males had the most extreme difference in coloration among the five color types. Suzanne found that blue males were most conspicuous to females, and had the highest mating success, in the shallow-water habitat where they were most abundant and also appeared most conspicuous to the human eye. Yellow males were more conspicuous and most successful in the root-spawning site, with its blue-green open water background, where they were also most abundant. We don't know yet if color vision varies between the environments, but this possibility has been investigated in Lake Victoria cichlids in the context of speciation.

Pundamilia nyererei, in which males are usually red, and *Pundamilia pundamilia*, in which males are usually blue, are two closely related species that were introduced earlier. Work has continued on this species complex, and provided insights into not only how environmental change can cause species to merge but also how environmental variation can contribute to the maintenance and formation of species through sensory drive. These two species often overlap in where they are found, but there are subtle differences as well, with *P. pundamilia* usually occurring in shallower water than *P. nyererei*. In locations with intermediate turbidity, Victoria's waters tend to filter short wavelength, blue light quickly with depth, resulting in *P. pundamilia* being found in a more broad-spectrum light environment and *P. nyererei* occurring in a habitat in which light is red shifted. This is thought to be the reason that the visual tuning of the two species also varies consistently with depth, as has been shown repeatedly along the descending slopes of different islands.

The opsin proteins found in the long wavelength, red-sensitive cones of *P. nyererei* have a genetic difference from those of *P. pundamilia* that causes their peak sensitivity to be a little further toward the red end of the spectrum, likely enabling *P. nyererei* to make better use of the limited light in their environment, much of which is red. This could help them, for example, in finding prey. Seehausen and his colleagues hypothesize that this difference also contributes to selection for red coloration in

P. nyererei males; not only do red males ensure they are reflecting light and making themselves visible in their local environment but they also ensure they are reflecting wavelengths of light readily detected by females. Of course, they may not contrast well with the background in terms of hue, but with modest ambient lighting there may be little option. Not all links in this hypothesized scenario are fully confirmed yet. Still, evidence of one key connection, linking survival to the light environment and thus natural selection, has now been obtained.

Martine Maan of the University of Groningen in the Netherlands led a study of how rearing in different light environments affects lab-bred *P. pundamilia*–like and *P. nyererei*–like (they are referred to as "like" because these fish came from Python Island and are the result of "parallel hybrid speciation"; henceforth I will omit the "like") as well as hybrids between them. Measuring survival was not initially a goal of the study; in their published paper describing the work, the authors note that "the results reported here emerged serendipitously from counts that were conducted for administrative purposes only." I am sure many scientists reading this feel envious of an important finding arising entirely as a by-product of another study and leading to a major publication. Many of our seemingly best-designed and most laborious experiments do not yield such consequential outcomes.

The key result was that the juvenile *P. pundamilia* survived better in a blue-shifted light environment, akin to that of the shallow waters they typically inhabit, whereas the *P. nyererei* suffered fewer deaths in more red-shifted light similar to what is present in the slightly deeper waters of Lake Victoria, where they are usually found. The differences were surprisingly large: almost 40 percent after a year. Hybrids of the two species were roughly intermediate, as would be expected. Hybrid survival did not differ between light treatments and from the "pure" forms in either light environment, although the trend was for them to generally do a little worse than *P. pundamilia* under blue light or *P. nyererei* under red light. There are at least two plausible reasons for the survival differences observed, though additional data would be needed to resolve which, if either, applies. First, fish may not be as good at detecting and capturing food in an ill-suited light environment. This could result in reduced growth, social disadvantages, and ultimately an early death. Alternatively, inappropriate light environments may cause stress and contribute to aggression and conflict. The essential outcome in terms of sensory drive and speciation is that sensory divergence that can influence mating preferences is shown to be favored by natural selection entirely outside the context of mating, facilitating the formation and maintenance of species. Such divergence could operate in isolation, but seems more likely to complement other forms of ecological divergence between environments.

LEARNING AND DEVELOPMENT

Biologists have often treated mating preferences and numerous other traits as if they were strictly genetically determined—as if, say, one allele means a female always chooses the red male while a different allele means she always chooses the blue male. It makes sense to start with simple assumptions like these, see how much insight they can provide, and then layer complexity on top. In the case of mate preferences, there is a good deal of layering that we can do. In particular, we know that experience and the environment in which an organism develops can modify sensory systems, color patterns, and social behavior, and such effects are now being studied and incorporated into our understanding of how species are formed.

Several studies have built on the Lake Victoria *Pundamilia* system. They were mainly conducted in Maan's laboratory in Groningen by a former research student of mine, Shane Wright. I sometimes tell Shane's story to my undergraduate students at East Carolina University, many of whom come from struggling small towns and cities in our economically stressed region and are the first in their families to attend university; Shane is a model for some. He too comes from a small city, in his case in western Virginia. He worked terrifically hard, funding his education himself, and made excellent use of the opportunities that came his way, including a spell in my laboratory working on stickleback fish. He went on to join the Maan laboratory for his PhD, conducting fieldwork in Tanzania and laboratory work in the Netherlands—all of it a long way from small-town Virginia. I visited him in Groningen while he was there and was reminded that he is one of the most positive people I have known.

Shane's investigations of how a fish's early light environment shapes its mate preference, vision, and color patterns made use of individuals from the rearing experiment described above in which fish were raised in different light regimes. When Shane and his colleagues tested the mating preferences of the surviving adults, they found that the early environment of the fish had had a significant effect. Fish raised in the shallow-water, broad-spectrum light regime in which *P. pundamilia* normally reside showed a stronger preference for blue *P. pundamilia* males than did fish raised in the red-tinged environment typical of *P. nyererei*; the latter individuals tended to respond more to the red *P. nyererei* males. In contrast, the rearing environment had little effect on how male color developed. *P. nyererei* males consistently turned out red while *P. pundamilia* males were consistently blue as breeding adults. Both, though, developed a little more green on their bodies in the red-shifted habitat.

One interpretation of the preference results is that an early light environment should have a similar effect to that of selection acting on vision and cause fish to prefer males native to the environment in which they were raised; this could contribute to speciation. But in the paper presenting these findings, Shane and his coauthors also point out that the opposite effect on speciation is possible, though perhaps less likely. Should a female end up in the "wrong" light habitat early in life, developmental effects will cause her to be more likely to choose males of that habitat rather than those of her own population—inhibiting speciation.

In additional analyses of the genes associated with color vision, there was no straightforward relationship between a female's opsin expression and her mate preference. The opsin genotypes (genotype refers to the variants of genes present at one or more genes) of the red cones, however, for which there was unexpectedly extensive mixing of different combinations of *P. pundamilia* and *P. nyererei* genes, proved more important. Under wide-spectrum light, females with exclusively *P.*

nyererei–like red cones preferred red males whereas females with only *P. pundamilia*–like red cones preferred blue males; females with one cone gene of each type showed no preference. Yet in red-tinged light, there were no clear differences in preference. The upshot of these experiments is that the light under which fish are reared affects mating preferences in the same direction—with deeper-water fish preferring red males and shallow-water fish preferring blue males—as the effects of the red cones characteristic of each depth.

During the rearing of the fish used in these studies, special care was taken to minimize interactions between the recently hatched young cichlids and their mothers. This is because the mother cichlid often cares for her offspring for several weeks, carrying them in her mouth, and then guards them even after they are free swimming—and parent-offspring interactions are well-known to influence mating preferences. Birds have been found to "imprint" on their parents and the songs they hear such that their choice of mates as adults reflects their experience during early life. Machteld Verzijden, then at Leiden University in the Netherlands, wondered if such effects might also be present in these cichlid fish and if they could play a role in the explosive speciation rates in the African Great Lakes.

To test this possibility, she switched fertilized eggs between mothers. Two to five days after spawning, she gently induced new *Pundamilia* mothers to spit out their recently fertilized eggs. She then moved the eggs between mothers—sometimes mothers of the same species, as a control, and sometimes those of different species. Mothers were left with the eggs and fry for four weeks, and then the fry were raised to maturity. The mature females were tested with a red male *P. nyererei* and blue male *P. pundamilia* using an experimental design that took advantage of the male-female size difference. Males were kept apart using grid dividers that the smaller females could cross but the males could not, ensuring females had a full opportunity to choose and males could not directly interfere with one another.

The results showed an unexpectedly strong effect of which type of mother had reared the females. Those reared by mothers of their own species preferred males of their own species. But females reared by females of the other species preferred those males! Thus, a *P. nyererei* female who had been mouth brooded and guarded by a female of her own species, but not her mother, preferred a red male of her own species over a blue male of *P. pundamilia*. But a *P. nyererei* female who had been mouth brooded and guarded by a *P. pundamilia* female rejected the courtship of males of her own species and responded positively to the courtship of *P. pundamilia* males. The key issue was the species of the female who had cared for her, not who her genetic parents were. The fish could interact fully, so it is possible that the preference was based on chemical cues, but other studies suggest these are not important in *Pundamilia*. Visual cues are a possibility because females, although generally similar between species, weakly express some aspects of the color patterns that distinguish males. In any case, what is unequivocal is that a female's choice of mate is influenced by the mother who rears her, and in a way that could accelerate speciation as populations diverge. It has long been thought that sexual selection can complement or even reinforce divergent natural selection's role in the speciation process if mating with members of one's own species is favored to avoid producing hybrids with poor prospects. These experiments suggest that imprinting may lead to an analogous outcome.

CONFLICT

We have seen how female preferences and male traits can coevolve to yield good "fits" within populations, and how early developmental effects and imprinting can complement such coevolution. Yet most of us have had our share of disputes with our family members (whether by birth or marriage) and friends. We know from experience that social interactions often involve conflict as well as cooperation, including the interactions involved in mating and reproduction. It is becoming clear that conflict is a major part of animal courtship as well, with potential consequences for speciation and diversification.

An example of sexual conflict is seen in Gray's work, on which I was a collaborator, on the fish *Telmatherina sarasinorum* in Lake Matano, Sulawesi. This species feeds mainly on the eggs of other *Telmatherina* as they are spawning, especially *Telmatherina antoniae*, but this feeding habit can lead to conflict in reproduction. When a female goes to spawn with a male, frequently other males rush in to try to fertilize the eggs too. The first conflict arises if the female is not interested in having the other males fertilize her eggs—but it gets worse.

Males sometimes turn around immediately after spawning and try to eat the eggs of the female they just mated with (figure 4.3). This rarely happens if there is only one male involved in the spawning. But as more males are involved, the probability of such cannibalism goes up rapidly.

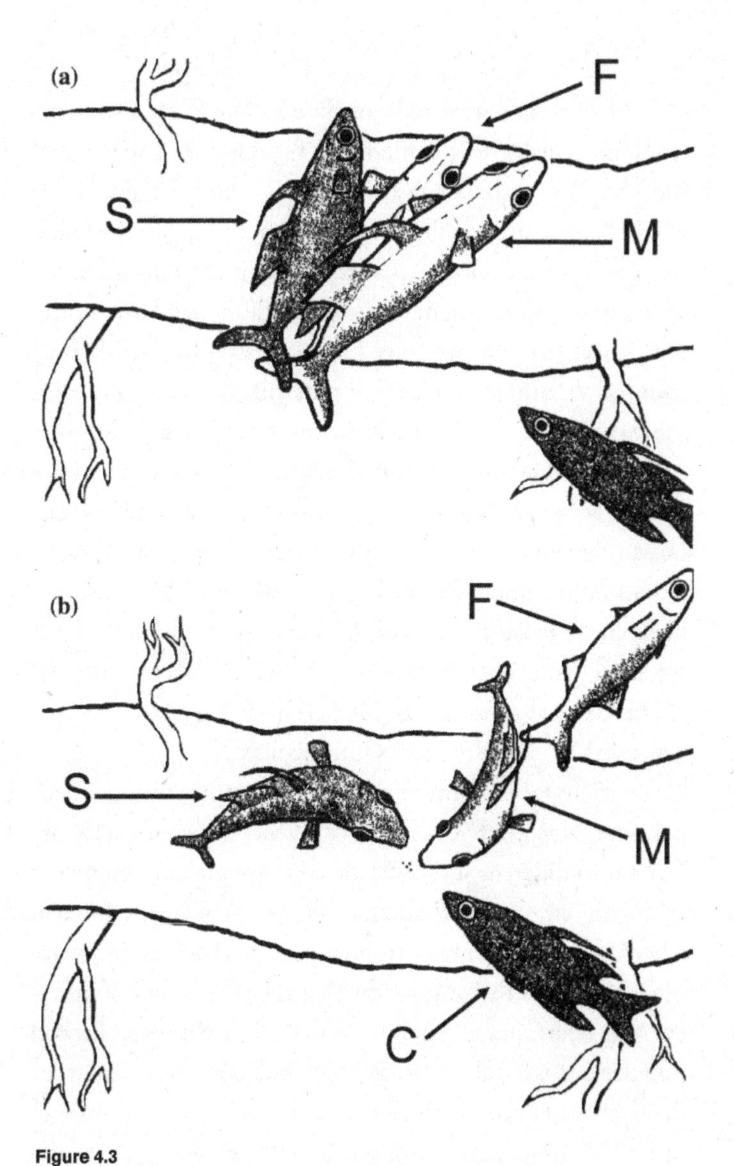

Figure 4.3

Spawning of *Telmatherina sarasinorum* sailfin silversides (a) followed by (b) males attempting to cannibalize eggs. F = female; M = paired male; S = sneaker male; and C = nonmating cannibal male. *Source:* Reprinted with minor modifications with permission from Springer Nature, from Gray and McKinnon, "A Comparative Description of Mating Behaviour in the Endemic Telmatherinid Fishes of Sulawesi's Malili Lakes," *Environmental Biology of Fishes* (2006).

When two or more additional males join in the spawning, the original male tries to cannibalize the eggs over 80 percent of the time. Before collecting any data, we hypothesized that this would happen because the likelihood that the male is the father should go down as more males are involved. There are just more sperm in the water around the eggs, and although Suzanne did not collect data on paternity for her study, it is well-known from other animal species that paternity is easily diluted as more sperm compete for fertilizations. In addition, if the other males are going to eat the eggs anyway, it probably pays for a male to try to get his share; if he can't become a father, at least he won't go hungry. Suzanne also collected data supporting the latter interpretation. Even with no sneakers, males were more likely to try to eat the eggs of the female with whom they just spawned when other males rushed in to feast. It was satisfying, if a little macabre, to see our predictions borne out. The paper was titled "Cuckoldry Incites Cannibalism."

A more subtle form of conflict can occur within individual genes when different forms of a gene, different alleles, are favored in different sexes. We already explored an example of such conflict, or sexual antagonism as it is formally known, when discussing the rapidly evolving sex determination systems of cichlids. It is becoming clear that the coexistence of different sex-determining systems even within cichlid species and populations is a quite ordinary state of affairs, and cichlid sex determination can rapidly evolve along different pathways in different populations. For instance, different genes can determine sex, or a gene for maleness can be dominant in one population (and femaleness recessive) and a gene for femaleness can be dominant in a different population (and maleness recessive).

As I write, at least twenty-two different sex determination systems have been identified in East African cichlid fishes, spread across eighteen of their twenty-three chromosomes. Potentially, such differences can cause incompatibilities between males and females of different populations and result in offspring of reduced fitness, thereby contributing to the evolution of reproductive isolation between the populations and ultimately speciation.

When crosses between young species have been conducted, the predicted incompatibilities have been observed. Sina Rometsch, Julián Torres-Dowdall, and Axel Meyer of the University of Konstanz compiled all available data on such crosses for cichlids. They confirmed that sex ratio distortions (unexpectedly high proportions of males, or of females) were common between even closely related forms and their frequency increased rapidly with the evolutionary distance between species—much more quickly than other incompatibilities, such as the reduced survival of hybrid offspring. Additional data and analyses will be needed to assess whether the surprising diversity of cichlid sex determination systems and the ensuing problems with some hybrids have helped cause the elevated speciation rates often seen in these fish—or alternatively, are themselves a by-product of explosive adaptive radiation.

* * *

Some traits are difficult to explain solely in terms of their survival advantage. For such traits, an advantage in mating is frequently the explanation and thus the process of sexual selection. Like natural selection, sexual selection may play an important role in speciation, and the direction and strength of sexual selection will also be influenced by an organism's ecology. Comparative evidence suggests that sexual selection has contributed to ancient lake adaptive radiations. Sexual selection necessarily involves a social component in which both learning and conflict may be important, with consequences for speciation. How natural and sexual selection interact to generate biodiversity is one of the most exciting areas of research in ancient lakes.

5 CRICKET, CABERS, AND THE SINISTER ADVANTAGE

Wallace did not anticipate the existence of this ancient lake system. If he had chosen a slightly different route over Sulawesi, maybe not the Galapagos finches, but the sailfin silversides of the Malili Lakes would have become the most popular reference system for speciation research.

—Fabian Herder and colleagues, 2006

Why do cricket and baseball matches usually feature so many left-handed batters, especially at the top levels and on the best teams? And what might the answer to that question reveal about the causes of biodiversity? I will start with a little background, and then come back to the intriguing issue of lefties in sports and the broader processes they illustrate.

Evolution requires genetic variation, which originates in mutations. Media reports often treat evolution and mutation as the same thing, but they are not. Mutations are, in essence, changes to an organism's DNA sequence. They matter for evolution if they occur in cell lines that can be passed to the next generation, usually via sperm, pollen, eggs, or another sort of reproductive cell. But most mutations are dead ends and lost shortly after they appear, even if they are in sperm or eggs. This is because random changes are usually disruptive, throwing a wrench into the finely tuned machinery of a genome with billions of years of evolution and selection behind it. These disruptive, harmful mutations are removed by natural selection that is "purifying." Purifying selection tends to reduce genetic variation or at least prevent it from increasing.

Yet some mutations are not purged by purifying selection. Some mutations stick around alongside other alleles, coexisting in the same population for the long term. The presence of such "standing genetic variation" in large numbers of genes provides critical raw material for evolution. It can speed up adaptation, speciation, and adaptive radiation a great deal—with a bigger tool kit along with a wider range of parts and pieces, innovation is easier and faster. For these reasons, any process that increases genetic variation in nature, or at least helps it persist, is important to understand for those investigating the causes of biodiversity.

A SINISTER SPORTING ADVANTAGE, BUT NOT IN EVERY SPORT

This brings us back to the issue of left- and right-handers in human populations. *Sinister*, by the way, is related to the Latin word for being left-handed; historically, being left-handed had negative associations, at least in some cultures. But to a biologist, the presence of both sorts of handedness is important and intriguing because it is such a conspicuous example of different forms of a trait coexisting, seemingly indefinitely.

Although left- and right-handers are present in most every human population, their frequencies are not always the same. One of the best-studied exceptions to typical frequencies is the elevated proportion of lefties in some sports (figure 5.1). Lefties are more common in cricket and baseball, as already noted, but there are also more left-handed fencers than there are left-handers in the general population. And the same goes for badminton, table tennis, rugby, soccer, volleyball, and ice hockey—as well as martial arts. Lefties are frequently more common than expected.

A hint at an explanation can be found by comparing these sports, all of which have interactive components somewhat like fighting, with those in which the contests are less direct. Less directly competitive sports that have been studied quantitatively include swimming, gymnastics, skiing, rowing, archery, ice dancing, and caber tossing—yes, there apparently are data on caber tossing. In these less directly interactive sports, the proportion of left-handers is about the same as in the general population at roughly 10 to 13 percent.

A powerful explanation for the difference between more and less directly interactive sports involves the costs and benefits of being rare. When a sport involves direct contests between individuals, a combatant/competitor can gain an advantage by having a strategy that is unfamiliar to opponents. For example, left-handed boxers, who are generally less common, will be familiar with right-handed opponents, as right-handers are frequently encountered. But right-handers will have less experience of southpaws—giving the rare left-handers an advantage. This is one illustration of a general principle known as

Figure 5.1

Top: In both caber tossing and swimming, there is no direct interaction between competitors and lefties have no advantage. *Bottom:* In fencing and cricket, left-handers are more common than in the general population. *Sources:* Reprinted with minor modifications, caber toss, public domain; swimming, E. J. Herson, Department of Defense news photos, under a Creative Commons CC BY 2.0 license (https://creativecommons.org/licenses/by/2.0); fencing, Marie-Lan Nguyen, under a CC BY 3.0 license (https://creativecommons.org/licenses/by/3.0); and cricket, Duncan Rawlinson, under a CC BY 2.0 license (https://creativecommons.org/licenses/by/2.0).

frequency dependence, which is when the success of a tactic or allele depends on its frequency, on how common it is. It is less likely to be an issue in a race between, say, swimmers, who never interact directly and simply must be fast, or for caber tossers, who need to be excellent at throwing logs end over end, but do so independently of other caber tossers.

Frequency dependence is important because if a trait or allele experiences greater success when it is rare, it will tend to persist in the population. This sort of frequency dependence is known, a bit more precisely, as *negative* frequency dependence because it involves a negative relationship between frequency and biological fitness. In such a scenario, as soon as the frequency of a trait declines for whatever reason, the success of the trait, its biological fitness, shifts in the opposite direction and increases. Individuals possessing the trait make a larger contribution to the next generation, leaving more offspring, and it starts to become more common again. But if its frequency goes up too much, its advantage disappears and other traits or alleles are favored. Thus, different forms of a trait are maintained in the population, doing especially well whenever they become uncommon, and variation thus persists or even increases. This explanation leaves out whether the allele is dominant or recessive, which can also affect the details, though it is a secondary issue.

Variation, especially standing genetic variation, is essential to evolution yet cannot be assumed. Other processes are constantly eroding variation—and not just purifying natural selection, the mechanism discussed earlier. Chance processes will trim away variation too. If populations are not large or fluctuate in size, there will be a continual process of random sampling with every round of survival and reproduction. Even as some individuals survive, mate, and contribute to the next generation, others will not, simply by chance. These chance effects, like lotteries, coin flips, or the dealing of cards, will cause alleles to drift randomly up and down in frequency. Sometimes, particularly in small populations, they will drift to the point that an allele is lost completely. It is the same principle that is in play when you pick cards randomly from a deck. Pick twenty cards and at least one will usually be a club. Pick two cards and you can easily find no clubs in your hand, entirely by chance. Hence the loss of alleles becomes inevitable if populations are small or enough time passes. All else being equal, then, genetic variation should be continually trimmed from most natural populations, unless some other process, like frequency dependence, pushes back.

In the case of human handedness, the process causing both lefties and right-handers to persist is probably negative frequency-dependent success in fighting. Sports act as a proxy for fighting, at least those sports with direct interactions between competitors. Conveniently for the researcher, there are immense archives of readily available sports data and statistics. These make it possible to confirm another expected pattern: that the frequency-dependent advantage of being left-handed should disappear once lefties are as common as right-handers and comprise 50 percent of the population. As expected, left-handed players do reach 50 percent in some interactive sports, but almost never surpass that level owing to their advantage fading away as they become ordinary and familiar.

Differences in the frequencies of left- and right-handers between males and females also provide support for the frequency-dependent combat hypothesis. Males in most cultures are more likely to be involved in fighting and combat, which can influence survival, mating success, and the ability to provide for offspring. More left-handed males than females are therefore expected, and indeed in large data sets there are usually a couple percent more lefties among males than among females.

Frequency dependence is not the only process that can maintain a balance among alleles in a population and preserve variation. Another important form of *balancing selection*, the formal term for selection that maintains different alleles by balancing their success, occurs when having different alleles for the two copies most organisms possess of each gene (being heterozygous) results in greater biological fitness than having two copies of the same allele (being homozygous). When having two different alleles is advantageous, neither allele is likely to disappear from a population and variation will be promoted. The most famous case study comes from humans: the allele that causes sickle cell anemia, if an individual possesses it exclusively (thus is homozygous), confers resistance to malaria when it is heterozygous and paired with a typical allele. This provides a survival advantage in malarial regions despite the suffering of those with the misfortune to be homozygous for the sickle cell allele. Genetic variation can also be maintained in a population if the success of different traits varies with habitat, time of year, or similar variables.

Some of the best-understood examples of balancing selection involve frequency dependence, and here ancient lake systems are making important contributions, with some captivating natural history mixed in. One of the classic examples has analogies with human *laterality*, the general term that includes left- versus right-handedness. I say classic because biologists have been talking about it for decades, but the story continues to take unexpected twists.

SCALE EATING, RIGHT OR LEFT

This case of laterality arises from a habit first reported, at least in a definitive way, in a 1954 paper with the title "A Curious Ecological 'Niche' among the Fishes of Lake Tanganyika." It was written by two scientists, Marlier and Leleup, based in Uvira on the northwest shore of Lake Tanganyika in then Belgian Congo. I cannot see mention of Belgian Congo without thinking of the many horrors associated with that place and era as well as the continuing suffering in the region; it is a set of associations completely at odds with a little paper about fish diets.

Marlier and Leleup report that in a group of Tanganyika cichlids, the Perissodini, the adults subsist mainly on the scales of other fish, which they tear off their living prey with fearsome teeth. Marlier and Leleup note that the fish they held in an aquarium would not eat "earthworms, fish powder or insects" or anything else they presented other than the scales of live fish. Although the report from Uvira does not comment on it, this feeding habit has since been found to be associated with laterality in both feeding behavior and more unusually mouth shape (figure 5.2).

One of the first publications to look carefully at evolution in these unusual fish had Karel Liem as the lead author, a fish biologist famous for his extroverted personality as well as his

Figure 5.2
Right- and left-skewed mouths of the scale-eating cichlid *Perissodus microlepis*, from above. *Source:* Reprinted with minor modifications, from Lee et al., "Genetic and Environmental Effects on the Morphological Asymmetry in the Scale-Eating Cichlid Fish, *Perissodus microlepis*," *Ecology and Evolution* (2015), under a Creative Commons CC BY 4.0 license (https://creativecommons.org/licenses/by/4.0).

many important contributions. Karel was one of my major advisers when I was a graduate student at Harvard, and I remain grateful to him. At that time, it was out of the ordinary for faculty hired into junior roles to receive tenure at Harvard—tenure was generally reserved for well-established senior scientists hired away from other institutions—and junior people typically left after several years, with their graduate students orphaned as a result. That happened to me with my initial advisers, Steve Austad and Bruce Waldman. Karel kindly took me into his lab, where I finished my thesis work. Also a gifted teacher, Karel was beloved by many students, and his introductory course was one of the most popular at the university. Yet he often did his best to make life difficult for administrators. If, for example, he was unhappy with the room he was given for a class, he would encourage students to have their parents call the university to complain about not getting their money's worth. Chairs and deans rarely saw the humor in this. Looking back, I have a better appreciation for Karel in a different way: for what a groundbreaker he was, and the challenges he must have faced as one of the first nonwhite faculty members in biology.

In their paper on scale eaters, Liem and his coauthor Stewart described the various adaptations to this feeding habit in jaws and teeth, in particular, and how the apparent extent of scale eating varies across the group. They described a new species with notably extreme laterality and suggested that adaptations in the head shape of members of this group arise mainly through the differential growth of skull elements. This proved to be an enduring insight.

The most influential paper on feeding and mouth laterality in Tanganyika's scale eaters appeared a little later and was written by Michio Hori, who was then based at Wakayama Medical College (Japan)—though like Marlier and Leleup, he conducted his research out of Uvira, which by that point, 1993, was part of the Republic of Zaire. For most of a decade, I presented the work described in this paper every time I taught a course in evolution to undergraduate biology students. It is quite literally a textbook case study.

Hori reported that his study animal, *Perissodus microlepis*, typically attacks by striking the prey fish on the side of the body, approaching from the rear to be less visible. Individual fish strike from mainly one side or the other, and their mouths are twisted correspondingly. This increases the area of the mouth and number of teeth in contact with the prey's body while still allowing the scale eater to sneak up from behind. For example, a fish that strikes from the right will tend to have a mouth that is twisted to the left (figure 5.2); conversely, a fish that strikes from the left should have a mouth that is twisted to the right. Prey fish of course seek to avoid having scales and sometimes pieces of underlying flesh ripped off.

If the scale eaters always struck from one side, say, the right, this would be simpler for the prey as they would only have to be guarding against *Perissodus* attacks from the right. Therefore in a population where attacks all came from the right, the occasional scale eater that bit from the left should do particularly well as prey should be paying no attention to that side. Of course as lefties became more common, the prey should shift their attention accordingly. In sum, the situation should lead to frequency dependence along with the persistent presence of both left- and right-attacking *Perissodus* in the population, at something close to fifty-fifty unless some other factor affects feeding success from each side.

Hori found that when he towed a prey fish behind a boat and captured wild scale eaters after they struck it, individuals attacking from the right always had a mouth with a left twist, whereas individuals attacking from the left always had a mouth that twisted right. He was also able to identify scales, from *Perissodus* stomachs, as having come from the left or right side of the prey based on pore patterns on the scales. Much as with strike observations, fish with right-twisted mouths had eaten scales from the left side of prey fish and left-twisted scale eaters had eaten scales from the right side of the prey. The most noteworthy part of the paper, however, was the evidence for frequency dependence.

Hori was able to test for frequency dependence by taking advantage of natural oscillations in the relative abundance of fish with right- or left-twisted mouths. The frequency of the two mouth types never got too far from fifty-fifty. But about every 2.5 years, the population shifted from mainly lefty fish, a maximum of about 65 percent, to mainly fish with right-twisted mouths, with a similar peak of about 65 percent. By looking at scars on the sides of prey fish, Hori was able to estimate the relative success of the morph that was less common at any given time. He found that as expected, whichever twist type was *less* abundant left more scars and achieved more successful scale bites, with success rates flipping when the rare morph's frequency increased and it became the abundant type. Looking at frequencies of each morph for breeding pairs possessing broods of offspring, he found that the probability of breeding was consistently the reverse of population frequency, much as for feeding success. Possibly because they were better fed, lefties were more likely to be breeding when individuals with right-twisted mouths were common and right twists were more likely to have broods of offspring when lefties were common (figure 5.3).

Comparing the direction of twisting in each member of a reproducing pair, relative to mouth twisting in the offspring they were brooding, Hori also obtained data with which to model the genetics of laterality. He reported that the frequencies of left versus right twists in the mouths of offspring could

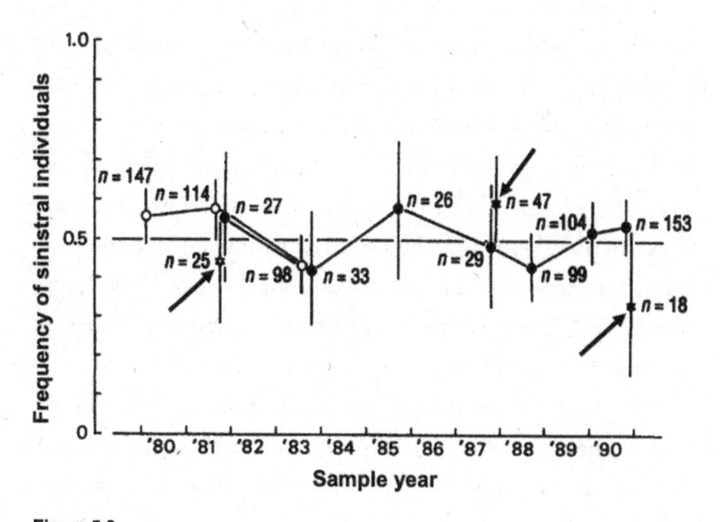

Figure 5.3

Michio Hori's (1993) original data showing oscillations in the frequency of individuals with left-twisted or sinistral mouths (the remainder being right twisted) for two sites–one open circles, and one closed. Stars, also highlighted with arrows, indicate the frequencies for breeding pairs, which are usually of the rare form. *Source:* Reprinted with minor modifications with permission from AAAS and Michio Hori, from Hori, "Frequency-Dependent Natural Selection in the Handedness of Scale-Eating Cichlid Fish," *Science* (1993).

be explained by a single gene, with a right twist genetically dominant over a left twist, though his sample size was modest. This genetic mechanism combined with a key detail of growth and development might help to explain the oscillations in frequency too. Since juvenile *Perissodus* are too small to eat scales, they feed on other prey for about two years before switching to a scale diet. Thus, any increase in the reproductive success of lefties, for example, does not affect the frequency of attacks by left-twisted scale eaters until the offspring are old and large enough to make such attacks, a couple of years later, potentially introducing a lag and contributing to oscillations.

There are few such well worked out examples of how frequency dependence can maintain variation in nature, so many biology instructors made use of this example in their teaching, as I had. Its appearance in textbooks encouraged that. Consequently, on a trip to Germany for a conference, I was interested to learn that there might be a good deal more nuance and complexity to the system. I was in Konstanz visiting my colleague Axel Meyer, along with a couple other evolutionary biologists, prior to the start of a meeting in nearby Tübingen.

Professor Meyer is a personality on the scale of Karel Liem (with whom he worked briefly) and has made seminal contributions to the study of ancient lake biodiversity. He is famous for the ambition, breadth, and productivity of his research program as well as his willingness to wade into controversial areas—sometimes upsetting an apple cart along the way. Axel is also a marvelous host, with a keen sense of humor. We stayed with Axel and his wife, Gabi, in their home on a hillside overlooking Lake Konstanz. Their house was full of oversize windows and striking views, and populated with large pieces of art and mementos, often with a whimsical or playful aspect.

While in Konstanz, I met with most members of the Meyer laboratory and learned that some were following up on Hori's work, looking at the development and genetics of mouth laterality in scale eaters. They had just started publishing a series of papers with Henrik Kusche and Hyuk Je Lee in the lead roles, and later Francesca Raffini. Their work, and that of other laboratories, raised the possibility that scale-eater mouths did not always twist as clearly to the left or right as had been thought, and while twisting was heritable, the genetic details could not yet be resolved. These studies, and continuing work by Hori

Figure 5.4
Axel Meyer and Ernst Mayr (discussed in chapter 2) in Konstanz, Germany, in 1998. *Source:* Axel Meyer.

and his colleagues, have also suggested interesting developmental interactions between the laterality of mouth twisting and the side of scale-eating attacks. Some results indicate that an early behavioral tendency to strongly favor attacking from one side, which could arise in various ways, could lead to a remodeling of the bones of the jaw and contribute to the twisting of the mouth. Bone is more dynamic than many of us realize, building up with use and disappearing with disuse, and in many organisms, side asymmetries in shape and form result from how each side happens to be used during development. This scenario would be consistent with Liem's suggestion that the growth of bones in the jaw was key to laterality. In addition, however, there is evidence of an effect in the opposite direction: the laterality of mouth twisting can influence behavior, and fish may learn to attack from the more advantageous side based on how their mouth twists. Effects in both directions are not necessarily mutually exclusive, and some differences in findings could be due to variation in the fish themselves, among populations. Research on the genetics and development of mouth twisting continues.

Studies to further test the possibility of negative frequency-dependent selection on scale eater laterality yielded more straightforward results than did the work on genetics and development. This research was led by Adrian Indermaur, working in the laboratory of Walter Salzburger at the University of Basel, Switzerland. I have met both scientists at ancient lake conferences as well as during a visit to Switzerland. Adrian seems the quintessential field biologist, bearded and unconcerned about the discomforts and dangers of the remote sites he often works in—to a degree that impresses even the other field-workers in his lab. Walter is also a dedicated field scientist as well as a remarkably productive scholar, who comes across as utterly unruffled by the dozens of projects that appear to be underway simultaneously in his group. His students talk about how envious they are of his ability to go directly to sleep on long bus and plane rides in Africa, so his good-natured demeanor seems an advantage during travel as well. When I visited his laboratory, I was much impressed by the cabinets of Tanganyika cichlids and related samples—a remarkably comprehensive set of males and females of almost every species in the lake, around 240 species in total.

Adrian, Walter, and two other Basel colleagues conducted a more definitive test of frequency dependence in *Perissodus* than in any previous study. They placed a series of underwater enclosures at six to nine meters depth on the bottom of Lake

Figure 5.5
Underwater enclosures used by Adrian Indermaur and colleagues to study scale eating, six to nine meters beneath the surface of Lake Tanganyika. *Source:* Reprinted with minor modifications with permission from John Wiley and Sons, from Indermaur et al., "Mouth Dimorphism in Scale-Eating Cichlid Fish from Lake Tanganyika Advances Individual Fitness," *Evolution* (2018).

Tanganyika, each enclosing sixteen cubic meters of water and open on the underside to the lake bottom, for realism. Each enclosure was stocked with prey fish, either 100 percent lefties, 100 percent right twists, or a fifty-fifty mixture of the two, though always the same total number of prey and scale eaters. The use of underwater enclosures is unusual in studies of lake fish—typically enclosures are in shallow water and open at the top—but the depth and location in the lake likely reduced disturbances by waves, larger creatures and passersby as well as providing suitable fish habitat (figure 5.5).

The frequency-dependence hypothesis predicts that prey fish can most effectively adjust to scale eaters of just one laterality, such as all lefties. With attacks coming predictably from one side, the prey should be able to avoid the scale eaters better than with a mixture of right- and left-biased *Perissodus*. Conversely, then, the *Perissodus* should experience the greatest feeding success in the mixed group, where the prey cannot predict the direction of attacks. This was indeed the case: by the end of the three-day experiment, prey fish housed with both lefties and right-twisted scale eaters had lost more scales on average, and the scale eaters were more likely to have scales in their digestive tracts. This was an important finding.

For a long time, a key gap in this body of work was a test of whether selection favors scale eaters with mouths that are strongly twisted, either to the left or right, rather than being just a little nonsymmetrical. This sort of selection is known as disruptive selection. It occurs when, say, a typical bell curve for some trait is "disrupted" by intermediate individuals having poorer survival or reproductive success, whereas individuals toward the extremes do well. Even evolutionary biologists have not always made the subtle distinction between frequency-dependent and disruptive selection, which can result from similar underlying causes, but it is crucial to investigate both.

Yuichi Takeuchi, Hori, and two colleagues made important progress on this missing puzzle piece. They looked for patterns indicating disruptive selection on scale eater laterality in a field study that extended across nine research seasons, from 1990 to 2014, all at the southern end of Lake Tanganyika. They quantified the number of scales in the stomachs of scale eaters as well as the strength of each individual's mouth asymmetry, and found that among similar-size fish, feeding success was indeed correlated with mouth asymmetry. Thus, fish with stronger mouth twisting were more successful in acquiring larger helpings of their grisly meals. It appears diversity is favored not only because bearers of whichever type of side bias, left or right, is less common are able to obtain more food but also because those with stronger bias eat the most. As a consequence, they should be better able to contribute offspring and their genes for extreme twisting to the next generation. Given our limited understanding of the genetic and developmental basis of scale eating, these results may not be the final word on disruptive selection in this system, but they are important.

SEXUAL TRICKERY

Biting scales off other fish is an implicitly sneaky way to make a living, but explicit duplicity (if that is not too much of an oxymoron) is also well-known in nature, once again with a frequency-dependent aspect. Theory predicts that deceptive individuals or behaviors will usually be rare in a population since otherwise the victims may learn to avoid tricksters, just as the prey of scale eaters can learn to which side to attend. And over the longer term, selection will more strongly favor the ability to detect deception. Some peculiar forms of deception are known from ancient lakes, with one of the most unusual involving the fish *Telmatherina sarasinorum*, a species studied by Suzanne Gray in Sulawesi's Lake Matano. We have discussed *T. sarasinorum* before with regard to courtship and reproduction; here we shift to foraging and feeding.

Over the course of three field seasons of work, Suzanne and her collaborators (I was involved too, though Suzanne and her assistants collected all the data) made over 130 separate observations of male *T. sarasinorum* feeding on the eggs of another common telmatherinid species, *T. antoniae*, at shallow beach sites. Individual *T. sarasinorum* males would follow a courting pair of *T. antoniae* as they swam along the mud and sand, with the pair occasionally laying eggs. Typical *Telmatherina* courtship is shown in figure 5.6. Each time the *T. antoniae* pair performed their spawning behavior, which involves pressing their bellies to the mud in unison and briefly quivering, the *T. sarasinorum* male would rush in and nip at the spot where the pair had just quivered, trying to eat the eggs. The *T. sarasinorum* males also did their best to monopolize their *T. antoniae* meal tickets. They fought often with other fish of their own species that tried to join in on the egg feast and in some instances even defended "their" *T. antoniae* pair from other male *T. antoniae* that were interested in challenging for the female. As I watched the *T. antoniae* pairs, I could not help but feel sympathy for them, constantly harassed by a hungry predator of their just-fertilized eggs as they tried to get on with their brief romance. Still, the situation sometimes became even worse for the little couples.

During four of their observations, Suzanne and her assistants saw the male *T. sarasinorum* chase away the male of the *T. antoniae* pair. This was surprising enough given that the pair was presumably feeding (unwittingly and one suspects unwillingly) the *T. sarasinorum* its eggs, but what came next was even more unexpected. The *T. sarasinorum* male took up where the departed *T. antoniae* male had left off, courting the female as she roamed over the mud bottom. When she settled to the substrate to spawn, the *T. sarasinorum* male accompanied her, though he did not quiver and presumably did not

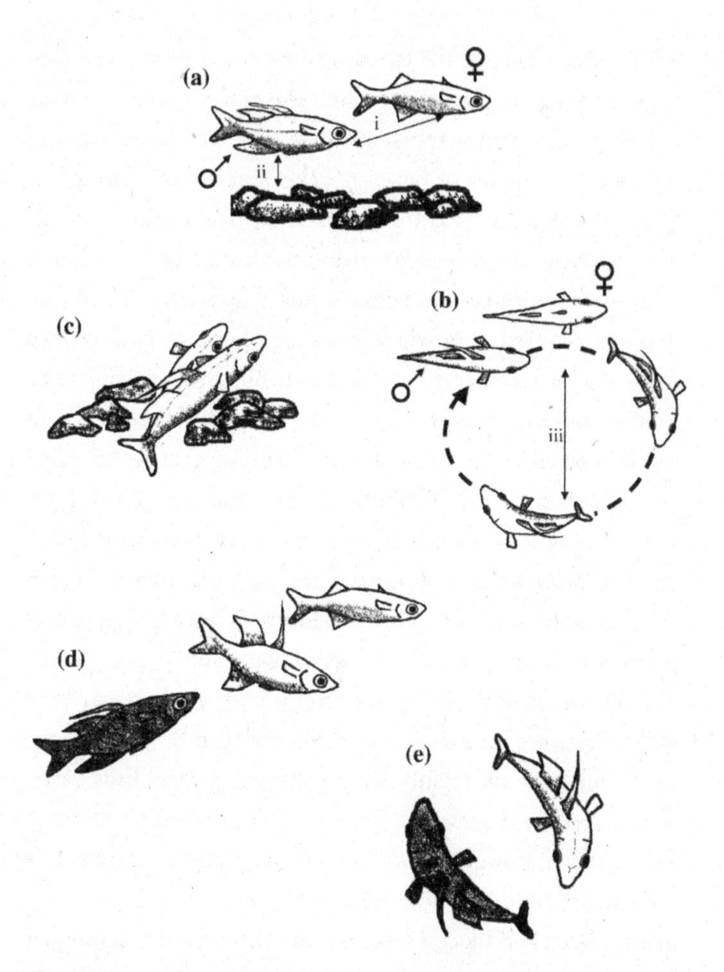

Figure 5.6
Typical courtship in the telmatherinid fish of Sulawesi. (a) Pairs swim together, (b) the male performs circles, (c) and the pairs lower to the mud and quiver together, spawning; males are sometimes challenged by other males of the same or (d) different color pattern, in some cases even species, and may (e) fight and raise fins in displays. *Source:* Reprinted with minor modifications with permission from Springer Nature, from Gray and McKinnon, "A Comparative Description of Mating Behaviour in the Endemic Telmatherinid Fishes of Sulawesi's Malili Lakes," *Environmental Biology of Fishes* (2006).

release sperm even as the female quivered and likely deposited eggs. The female then moved on, as is normal, but the male turned around and snapped at the mud where the female had just been in pursuit of her eggs. This "sneaky egg eating," as Suzanne termed it, was the first case we could find of a male of one species deceptively courting the female of another and then eating her just-laid eggs. Other instances are known of deceptive courtship in which a female of one species courts a male of a related species and then eats him, famously in some fireflies, but eating eggs seems to be unusual. Observations of this odd behavior have also been made by another group, Alexander Cerwenka and collaborators, so Suzanne's observations were not just a freak occurrence in one or two eccentric fish.

It is nonetheless noteworthy and possibly important that this behavior was observed so seldom. If sneaky egg eating were happening at a high frequency, we would expect female *T. antoniae* to either learn to detect deceptive *T. sarasinorum* males or evolve an aversion to these males in response to the strong selection that results from many eaten eggs. Thus, rarity is expected. A proper test for frequency-dependent selection on this behavior, however, would require an experiment more like Indermaur's with the scale eaters, in which frequencies were manipulated and success was tracked. That would be a tougher order for sneaky egg eating, which is performed by males only now and then, and not known to be associated with some other obvious trait, in contrast to the strong association between the direction of scale eating and mouth twisting.

These accounts of Suzanne's work would be incomplete without describing the context for these fascinating observations. On the one hand, working in the Malili Lakes was quite comfortable as fieldwork goes because INCO, the company that owned the nearby mine (where my relatives had worked), considerably facilitated our research. INCO usually provided us with comfortable housing and food along with a cheap rental car. On the other hand, there was rather disturbing and frightening terrorist activity in the region when Suzanne was working in the 2000s. As her project was getting underway, the Al Qaeda–affiliated group Jemaah Islamiyah set off a series of bombs in nearby Bali that killed over 200 people, many of them Western tourists, as well as staging attacks in Jakarta and elsewhere. Muslim-Christian conflicts were taking place around the same time in nearby Poso (a town near the eponymous lake, a few hundred kilometers to the west of the Malili Lakes), and Maluku (aka the Moluccas), the islands just east of Sulawesi. The Poso and Maluku attacks resulted in the deaths of hundreds of people. At an especially tense point, physically huge and terrifically intimidating security officers, some reportedly with backgrounds in Britain's Special Air Service, started to appear around INCO and Sorowako, the town where we were based. Although there were no terrorist attacks in Sorowako while we were there, we later learned that a cell had been conducting training exercises in a nearby forest area.

I well recall the difficult discussions we all had about whether it was just too dangerous for Suzanne or any of us to continue our work. For myself, I thought sometimes of my master's thesis research, which had taken place along the coast of Peru when the Sendero Luminoso, a Maoist revolutionary group, was active there, and parts of the country were in near civil war. We had to be careful, but both the work and our research team turned out all right. Suzanne had the further challenge, especially when she was there on her own, of trying to run a research program in a somewhat remote locale while working overwhelmingly with men, some more progressive than others. But Suzanne was brave, in most every sense, and determined; I am still impressed at how she persevered and collected an exceptional body of data. I wish I could say our story was unique, but field biologists have often had to deal with political conflicts and even civil wars while doing their work, in addition to the serious risks frequently posed by tropical disease, jammed roads, and remote research sites. As Westerners, we were of course fortunate to be able to fly away to safe communities at the end of our fieldwork, in contrast with some of our Indonesian and, in my earlier studies, Peruvian colleagues. One of the more heartbreaking pieces of news I have received from a colleague was from Indonesia, when a friend and collaborator lost a small child during a dengue fever outbreak.

COLOR, CONTRAST, COURTSHIP, AND CONFLICT

I earlier discussed another finding of Suzanne's from her work in Lake Matano: at different sites around the lake, the male *T. sarasinorum* color pattern that contrasted most strongly with the background tended to be favored and more abundant. If movement between sites was infrequent, and thus the flow of genes weak, it is not hard to imagine different color patterns becoming established at locations with contrasting backgrounds—with some variation at each as a result of occasional migrants (and the migrants' descendants) moving about the lake. But thinking about color and contrast led me to wonder if such microhabitat-specific advantages, even occurring over modest distances and with frequent movement of individual fish between habitats, might still lead to an advantage for rare male color patterns and tend to promote a more variable population. To address such a question requires a theoretical analysis. I lacked the training to do the necessary mathematics, but I was fortunate to know a then up-and-coming theoretician, Maria Servedio, who I thought might be able to help.

I knew Maria because I had been the graduate teaching assistant in an animal behavior course she took as an undergraduate, which eventually led her to a position as a summer research assistant on some of my PhD studies with stickleback fish. We kept in touch as Maria carried on with her training and career. She matured into a leading theorist and took a position at the University of North Carolina, just up the road from my current location.

Maria brought in a talented PhD student, Amanda Chunco, to lead the modeling, which confirmed my suspicion of rare male mating advantages, at least if some plausible assumptions are satisfied. In the model, females prefer the most conspicuous male—let's say blue or yellow in color—the habitats (and backgrounds) that make either male type appear most conspicuous are about equally common, and females are evenly distributed among these habitats. With these conditions, males of whichever color pattern is rare will tend to obtain more matings. Basically, if yellow males get almost all the matings in the habitat where they have higher contrast and yellow males are also uncommon, they will get more matings per individual than the abundant blue males do in the habitat where blues do better. If the situation changes and yellows become more common, the advantage shifts and now rare blue males do better per capita. Greater success when rare helps both yellow and blue males persist even if the relative number of one-color type experiences a chance dip—as can easily happen.

There are other ways that female mating preferences can lead to the frequency dependence of male mating success and help to maintain variation. A quite distinct and intriguing behavioral process that occurs within a sex may also be important, however, and has emerged mainly from studies of ancient lake cichlids. This involves conflict and aggressive interactions, between males or females, that depend on color pattern.

Negative frequency-dependent selection may arise through aggressive interactions if males (usually the sex studied in this context, but females too in some cases) direct more aggression toward individuals who possess the same color pattern or are similar in some other way. Considering once more the example of blue and yellow fish, the idea is that blues are more aggressive toward blues and yellows toward yellows. This pattern could arise in several ways, but we will simply focus on what happens if it does. If blues become more common, then each blue male will experience more aggression from the abundant fish that share his color pattern and direct their belligerence toward him. In this scenario, uncommon yellow males get little grief from the ubiquitous blues, who tend to ignore them. Yellow males may seek disputes with other same-color males, but the uncommon yellows seldom bump into one another, and hence are spared a good deal of hostility and conflict. If the frequencies reverse, yellow males will be the ones perpetually engaged in disputes while blue males will go unmolested. Continually defending oneself from aggression is likely to be stressful, leaving little time or resources for feeding, courting, or other activities associated with high biological fitness. The stress can even lead to illness, as any overburdened parent, student or employee can attest to after ending up with a cold during a period of high tension and reduced rest. Through this mechanism, rare types of males will have increased fitness and two or more male forms will be expected to persist.

This idea was first suggested by two scientists we have met before, Ole Seehausen and Dolph Schluter, based on a collaboration they pursued when Ole made an extended visit to Vancouver, Canada, where Dolph is located. In their initial efforts to test it, they looked at patterns in cichlid territories across a set of relatively isolated sites in Lake Victoria in which males hold territories for courtship and mating. Cichlids defend their territories from other males, sometimes of other species as well as males of their own. Seehausen and Schluter examined several species simultaneously, some closely related and differing mainly in color pattern, and others more distantly related and differing in several types of traits. They suspected that color might influence aggressive interactions and the distribution of territories between closely related species (some still in the process of speciation) in addition to within species, and that negative frequency-dependent selection acting on color could thus play a role in speciation and species persistence as well as maintaining variation within a species.

According to the frequency dependence hypothesis, males from closely related species, which are most likely to interact, should typically differ in color when present at the same site because males with different color patterns more easily coexist. This is the very pattern Seehausen and Schluter observed in their data. Yet it was not observed for distantly related species that interact less intensely and frequently, again as expected. Similarly, when looking at individual territories rather than whole populations, they found that closely related neighboring males belonged disproportionately often to different species with different color patterns. This is the pattern expected if more similar males target each other for aggression, making it more difficult to live side by side.

Other hypotheses have been suggested that could explain some of the patterns documented by Seehausen and Schluter, so testing this idea in multiple ways is also important, such as by looking directly at the intensity of aggression between males with the same or different color patterns. Several such studies have been conducted, and they have often found evidence of the predicted pattern of stronger aggression between more similar individuals—but frequently with complications. For example, males with one color pattern might show the expected bias and males with a different pattern fail to, or an experiment may only partially test the hypothesis by looking at biases in just one type of male. At present, it appears that biased aggression plays a role in maintaining variation within some ancient lake species and may contribute to the formation and persistence of species. Many times, though, additional traits and processes are also involved in maintaining variation.

* * *

Genetic variation is an essential raw material of evolution, including speciation and adaptive radiation. The astonishing range of forms and traits present in most every natural population is fascinating in its own right as well. Ancient lake studies have provided important instances of a key process that can maintain biological variation, negative frequency-dependent selection, in which the long-term fitness of different forms of a trait are balanced. Frequency dependence can arise in diverse ways, including scale eating by cichlids with mouths twisted to the left or right, and more frequent aggressive interactions between fish with similar color patterns. What unites these examples is interactions between individuals such that those possessing the less common form of a trait are more successful.

6 ON THE VIRTUES OF INTERSPECIFIC RELATIONS

Wavuvi wazee husema kuna nyoka mkubwa sana katika ziwa, ana macho mekundu yapata kama futi 50. Huyu nyoka huibuka juu baada ya miaka 50. Akiibuka tu, wavuvi huvua sana samaki.

The old fishermen say there is a huge snake, which has red eyes and only comes up every 50 years. When it comes up, fishermen catch a lot of fish.

—Traditional story (translated from Kiswahili) of Lake Victoria from Mbarika, Mwanza Gulf, Tanzania

A birder trying to add gulls to their life list will often need a good deal of patience, persistence, and tolerance of failure. The same goes for a hiker trying to identify the oak trees they happen across while walking through a forest in the eastern United States. The problem is that many individual gulls and oaks can never be identified to the species level, even with a specimen in hand and DNA sequence, let alone from a brief glance. This is because gulls of different species frequently mate with one another, yielding offspring that don't belong definitively to either parental species. Oaks do the same. Sometimes these hybrid offspring go on to mate with other hybrids, or they may "backcross," mating with individuals of one of the parental species; things quickly become messy.

Beyond birders and hikers, the issue of interspecific hybridization and gene transfer between species (technically known as introgression, different from hybridization in that it emphasizes the transfer of genes rather than the matings themselves) have attracted more widespread attention in recent years as evidence has emerged that billions of us are carrying DNA that originated with Neanderthals and Denisovans—which are generally considered different species from ours. Moreover, some of these introgressed genes can influence important traits such as how likely we are to contract serious diseases, and how such diseases affect us. For example, two different fragments of the Neanderthal genome, which many people possess, have been shown to be correlated with how severely ill a person is likely to become if infected with SARS-CoV-2, which causes COVID-19. On the other side of the disease coin, the exchange of genes between different strains of influenza has resulted in some of the worst flu pandemics.

These observations clearly pose a challenge to the biological species concept, and the issue is not new. Nevertheless, hybridization was long treated as anomalous by the mainstream of science and quietly neglected. With more and more genomes becoming available, including large samples of genomes from different individuals for some species, hybridization and introgression have become much easier to detect and study, and their evolutionary significance has become more apparent.

One consequence of extensive hybridization can be extinction. We explored earlier how human-caused changes to lake water and light transmission can cause formerly well-defined cichlid species to collapse into a single gene pool. This can also result when people move animals into new locations where the new arrivals encounter populations or species with which they can interbreed, but from which they were previously separated by distance or physical barriers. For example, nonnative tilapia (*Oreochromis* sp.), escaping from aquaculture or released deliberately, with usually good intentions, are prone to mating with native wild fish and thus are putting native species at risk in some Tanzanian waters. The problem can be worse if the environment is disturbed in a way that makes interbreeding more likely or helps the hybrid offspring survive.

The effects of hybridization are not always so harmful, however. Hybridization can at times contribute to speciation, adaptation, and even adaptive radiation. In plants, the origin of new species through hybridization, or hybrid speciation, is surprisingly common, but involves genetic mechanisms that occur less frequently in animals. Nevertheless, young animal species with clearly hybrid origins have on occasion been documented. In chapter 2, we explored examples from ricefish and cichlids in which hybridization was involved in the formation of new species, including parallel hybrid speciation in *Pundamilia* from Lake Victoria. An additional pair of likely cases comes from Lake Victoria, this time involving "parental" species so distinct that they are in different genera—*Pundamilia*, once more, and the related genus, *Mbipia*.

Irene Keller and her colleagues in Ole Seehausen's laboratory at the University of Bern collected genomic and other data from five members of these genera at a single site in Lake Victoria, Makobe Island. Their analyses suggest that two of the species, one in each genus, are a result of recent hybridization between a species of *Pundamilia* and species of *Mbipia*. With such a messy situation, the existing classification scheme and genus names are in serious trouble, although to be fair, classification is generally a fraught enterprise in Lake Victoria, certainly when attempted above the level of species.

Hybridization can also contribute to adaptation without any need to form a new species. The steps and mechanisms involved may be straightforward or more subtle. In the simplest scenario, matings between species may enable one partner lineage to acquire a beneficial allele from the other. The benefit of the "imported" allele may, for example, be that it can replace a gene damaged by mutation. Or the new allele may be better suited to a changed environment; adaptation to higher temperatures comes to mind in our time. Introgression of this sort certainly occurs and may contribute to a commonly observed anomaly: analyses based on mitochondrial DNA often indicate different evolutionary relationships for a set of species than do analyses based on the remainder of the genome, which is generally referred to as the nuclear genome because it is found in the nucleus.

Mitochondria are key organelles in the cells of animals and plants, performing certain valuable metabolic reactions that cannot be completed without them. Mitochondrial DNA is separate from the rest of the cell's DNA because mitochondria evolved from bacteria that were engulfed by our ancient single-celled ancestors. Over time, mitochondria came to live permanently inside our cells in a mutually beneficial arrangement: we got a more efficient metabolism, and they got a safe home stocked with all of their needs. Thus, each animal, each of us, is the result of a symbiosis, which means organisms of different species living together. The organization, structure, and seeming unity of our bodies is a bit illusory—the result of an ongoing collaboration between two distinct genomes. This is one of those familiar facts of biology that can become remarkable again with a moment or two of reflection.

Housed inside the mitochondria themselves, mitochondrial genes are inherited almost exclusively through eggs (in animals), separately from an organism's other genes. Essentially clonal, mitochondrial genes do not take part in sexual reproduction. With these differences, it is not surprising that when we estimate evolutionary trees using the mitochondrial genes from a set of species or populations, the results sometimes differ from the trees suggested by the rest of the genome. Occasionally, though, the inconsistencies are major. Different species or subspecies sometimes share nearly identical versions of their mitochondrial genes despite substantial differences across the rest of the genome or for more visible traits. These inconsistencies can be explained by hybridization that is followed by selection favoring the new mitochondrial type, allowing it to sweep through the population. This may occur even if little mixing happens between the species for the rest of the genome. Selection is not the only explanation for such patterns, and more haphazard processes can also result in mitochondrial sweeps. For our purposes, it is the resulting discrepancy that is mainly of interest. These discordant patterns between mitochondria and the rest of the genome, which have been seen in several ancient lake radiations such as Sulawesi's telmatherinid fishes, were one of the first indications that hybridization may be more common during adaptive radiations than had been suspected.

THE TRANSGRESSIONS OF GENES

The newly mixed genes of two species can sometimes produce traits quite different from those present in either parent, and not simply in the sense of being intermediate or even combining traits previously found only in different species. On the surface, the parental species in this scenario may be outwardly similar to each other. Naively, we might expect that their hybrid offspring would be similar to both. Suppose, for example, that two species of lemur had evolved an elongated finger to help them catch insect larvae living in holes in trees (lemurs do such things, but the details in my example are hypothetical). If the superficially similar parental species mated, their hybrid offspring might be expected to have similarly long fingers, and if the hybrids continued breeding with one another we might expect things to stay about the same, generation after generation.

But depending on the genetics and details of how the long fingers of each species develop, things might be more complicated and more interesting. For instance, one species might achieve long fingers owing to a mutation favored by selection that causes an extra bone to be added. This is plausible since primate fingers already vary in the number of bones they possess, with the human thumb, for example, having lost one of the bones present in the other fingers.

In the second species with a particularly long finger, a single finger bone might grow longer, with no change in the number of bones. When genes for these different mechanisms end up together in hybrid offspring, the resulting finger lengths could be more extreme than those of either parent, depending

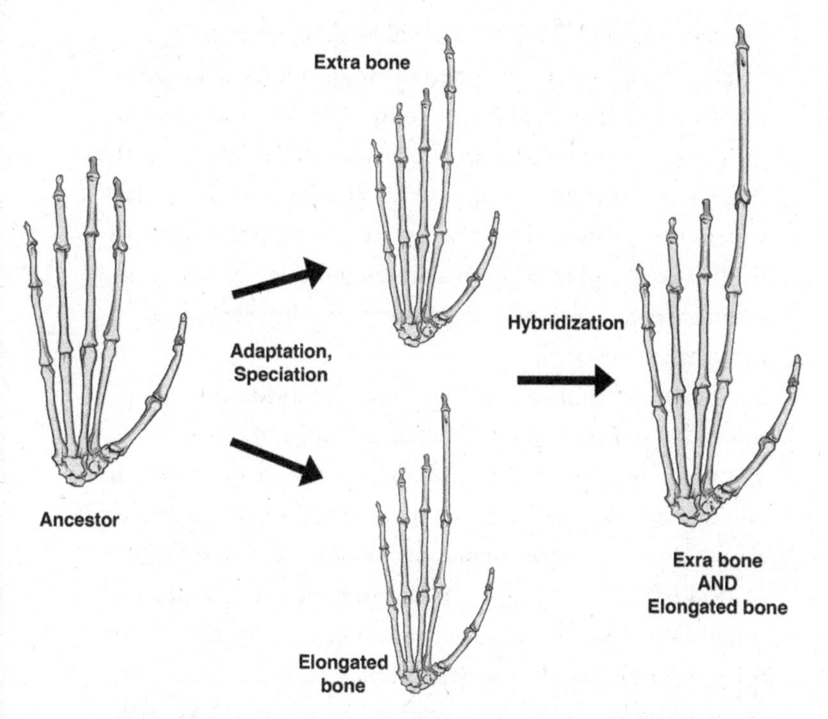

Figure 6.1

An ancestral lemur gives rise in different places to two new species, each having evolved a long finger to catch insect larvae in tree holes, but via different genetic changes. When they hybridize, the hybrid has a longer finger than either parent. *Source:* Haleigh Mooring.

on which alleles are dominant or recessive. For example, if both an elongated bone and the addition of a bone to each finger are dominant traits, hybrid offspring will have one notably long finger bone as well as an extra bone (figure 6.1). With two mechanisms of finger elongation both present in individual hybrid lemurs, those hybrids should have even longer fingers than either of the parental species—which already had impressive digits. The outcome will be different if the genetic details are different, but as long as there are not too many genes involved and the hybrids keep interbreeding for a few generations, we expect to see more extreme finger lengths in some of these hybrid individuals (figure 6.1). The key point is that the variation in the hybrids is expected to *transgress* the range seen in the pure (nonhybrid) forms, leading to this phenomenon's technical name, *transgressive segregation*, or a little less formally, *transgressive variation*.

The opportunities for transgressive variation can be surprisingly extensive. In our imaginary lemurs, there are other ways fingers could elongate. Considering just one bone, it could end up longer by either growing faster or for a longer period—processes that could easily involve different genes. Or instead of the final finger bone lengthening, it could be the middle bone that elongates or the first bone at the base of the finger. Combining these different bones with the alternatives of faster- or longer-duration growth, the possibilities start to add up fast. And we have not even thought about what could happen at the levels of cells, molecules, and biochemical pathways. In addition, transgressive variation can arise through other processes subtly different from the example here. There are a lot of possibilities.

Conceptual illustrations are helpful for thinking through and explaining the possibilities, but what evidence is there that transgressive variation has contributed to evolution? One of the first and most influential studies in this area, helping set the stage for a long series of ancient lake studies, involved fruit flies. It was conducted by a giant of population genetics and one of its more memorable personalities, R. C. "Dick" Lewontin.

I knew him a little bit because I took his class in population genetics while a PhD student. He left a strong impression. Lewontin was famously a Marxist and interested in student learning but disinterested in grades, so he let us know at the course's start that the default plan was for everyone to get a B+. A charismatic and slightly eccentric lecturer, he almost always dressed in a near uniform of denim shirt and khaki pants, and addressed the class with some of the quirky mannerisms you might include if you were designing an old-school professor from scratch. He was also generous with students, at one point initiating a weekly reading group, almost a course, in an area he was not enthusiastic about because students were interested.

Lewontin's seminal publication on the role of hybridization in adaptation continues to be cited annually in dozens of scientific papers, including articles on ancient lake radiations, more than fifty years after it first appeared. Together with collaborator L. C. "Charles" Birch, he studied a late nineteenth- and twentieth-century expansion in the geographic range and temperature tolerance of an Australian fruit fly, *Bactrocera tryoni* (then called *Dacus tryoni*), an important agricultural pest. Based on observations of apparent hybrids in nature, they inferred that interbreeding with a sister species had occurred, possibly as a result of human changes to their habitats. They hypothesized that the introgression of genes into *B. tryoni* had facilitated adaptation to a new temperature range not formerly tolerated by either species.

To test important assumptions of this hypothesis, they collected the two species from their natural distribution and started hybrid and pure *B. tryoni* populations in the laboratory at a range of temperatures. The flies are small and short-lived enough that one can use them to study the process of evolution experimentally, allowing different populations to evolve over several generations. Initially the hybrids did not survive and reproduce as well as the pure populations, but as time passed the hybrids improved, and also came to resemble the *B. tryoni* in appearance. At moderate, relatively optimal temperatures, the later generations of hybrids roughly matched the performance of the pure *B. tryoni.* But at the most extreme temperature, 31.5°C, the hybrids were superior. This is a remarkable finding. That temperature is clearly beyond the optimal range for the sister species as well as *B. tryoni.* Somehow the combination of genes from the two species substantially improved survival and reproduction *beyond* the usual tolerance of either "parental" species. In this system, therefore, there was transgressive variation for biological fitness at high temperatures.

Such multigenerational experiments have not been performed yet with ancient lake species, but a notably original study of the performance of hybrids in novel ecological situations was conducted by Oliver Selz. Working with Seehausen in Kastanienbaum, he conducted this "proof-of-concept" experiment using African cichlids. I met Oliver before he performed these experiments, when he visited my laboratory early in his graduate studies to discuss a possible internship. Even just starting his PhD, he was an impressive scholar as well as delightful guest (even bringing fondue ingredients with him from his native Switzerland).

The transgressive segregation study began with Selz and Seehausen setting up matings between a Lake Victoria *Pundamilia* cichlid and two other species, each with a different ecology. The *Pundamilia* sp. had a narrow, specialized diet of zooplankton, the tiny crustaceans and other animals that circulate in the open water of the lake. One of the species with which the *Pundamilia* were mated was a specialist feeder that scrapes algae from rocks; it is also native to Lake Victoria. The third species used in the crosses was a generalist, with broad, flexible feeding habits and tolerances, and native to Lake Malawi. The familiar part of the study involved measuring the feeding success of each species and their hybrid offspring on the prey types typical of those eaten by the two specialist species. The ecological speciation hypothesis predicts that the specialists should be most successful on the diets to which each has been adapting for many generations, compared to the other pure species or the hybrids. The more noteworthy part of the study involved presenting the pure parentals and hybrids with food items distinct from those of the parental specialist species. The novel foods were a freshwater shrimp and gammarid crustacean (the latter distantly related to the gammarids that radiated so spectacularly in Baikal), which are similar to prey eaten by some other species of fish, including other cichlids. If hybridization and transgressive variation give rise to novel forms with the ability to colonize new ecological niches, the hybrids should be superior on at least some of the novel prey. It was a potentially risky test of a bold hypothesis.

In their initial analyses, Selz and Seehausen confirmed the long-standing prediction that pures, when feeding on their usual prey items, should generally outperform hybrids and specialists adapted to other prey. For example, algae scrapers fed more successfully on algae (actually a laboratory approximation) than did the zooplankton feeders or the hybrids. Thus, the baseline results were largely as expected. But on novel prey, things were more interesting. On the unfamiliar prey items, the hybrids always did at least as well as the parental pures, and in some cases they did better. Specifically, the hybrid offspring of the algae scrapers and zooplankton feeders were the best at catching a novel and ecologically distinct prey species—freshwater shrimp. Similarly, in a different crossing in which the zooplankton eater, the generalist, and their hybrid offspring were each given an alga-like prey item, quite unlike either parental's usual foods, the hybrids again had the greatest success. The authors interpret their results cautiously and present a number of caveats. Yet their evidence that hybrids can surpass their parents in some novel ecological contexts argues that hybridization, possibly involving transgressive variation, has at least the potential to play an important role in adaptive radiation.

Speciation typically involves barriers to mating too. In a related study, Selz and several collaborators asked if hybridization might initiate such mating isolation. They asked if hybrids between cichlid species might sometimes mate more often with each other, preferring the same sort of hybrids, than with either of the parental species. As in the feeding study, they again began by crossing a set of Lakes Victoria and Malawi cichlid species with each other, generating sets of hybrid offspring from three pairings of parental species. These were raised apart from their mothers to avoid imprinting effects. When females of each of the parental species were allowed to choose among males of their own species, male hybrids, or males of a different species, they always showed a strong preference for males of their own species, mating with them almost exclusively. This was not a surprise, but much more difficult to predict was what the hybrid females would do. One commonality was that all three sets of hybrid females rejected the males of at least one parental species, and all mated fairly readily with hybrid males. The most startling and intriguing finding, though, was that hybrid females of a pair of parental species from Lake Malawi strongly preferred hybrid males, rejecting both types of "pure" male. Hence partial premating isolation arose in a single generation between hybrids and their parental species, and nearly complete premating isolation in some instances. When Selz and colleagues analyzed color variation in the Malawi pair, their results suggested that hybrids were transgressive for color. And since the transgressive color patterns were preferred by hybrid females but rejected by pures, it follows that the preference could also be considered transgressive.

Even with strong premating isolation between the first generation of hybrids and parental species, it is unlikely that a new species would be instantly established by the hybrids. The problem is that first-generation hybrids do not usually "breed true," to use the language of animal breeders, showing the same traits one generation after another. Rather, when the hybrids interbreed, their offspring will generally exhibit more of the variation present in the pures, with different combinations in each individual second-generation hybrid. Still, if both sexual and natural selection were acting to favor hybrid traits, a new species could emerge over time.

In a quite different but notably detailed example from a small radiation, the *Cyprinodon* pupfish of the Bahamian island of San Salvador, hybridization and introgression are hypothesized to have facilitated both ecological innovation and speciation. Although not located in an ancient lake, these pupfish have strong links to ancient lake systems. They are members of the same family that radiated in Lake Titicaca and include a scale eater, reminiscent of Lake Tanganyika, but in this case with apparently no right or left lateral twisting.

Emilie Richards, Chris Martin, and colleagues, working mainly at the University of North Carolina and University of California, have found that different populations and species of *Cyprinodon*, distributed across the Caribbean and nearby landmasses, almost invariably have a simple diet of algae and detritus (i.e., bottom muck). However, in San Salvador's lakes, which are less than about 10,000 years old, a species of scale-eating specialist (or even two) has evolved as well as a snail-eating specialist. Overall, the San Salvador pupfish miniradiation is not especially genetically variable, but is noteworthy for possessing unusually large amounts of adaptive genetic material from distant Caribbean islands—though from generalist species—and these genes are disproportionately associated with the newly evolved specialist feeding habits. Thus, "ancient" alleles present in generalists from different localities have been reassembled through hybridization, introgression, and natural selection into new adaptive combinations, reminiscent of the fruit flies studied by Lewontin and Birch, though more extreme in their innovations.

HYBRIDIZATION AND INTROGRESSION COULD HAVE ACCELERATED ADAPTIVE RADIATION. . . . DID THEY?

Several lines of evidence suggest that ecological opportunity, combined with the capacity to evolve reproductive isolation quickly through sexual selection, are important for explaining the most extreme ancient lake radiations such as the cichlids of the African Great Lakes. But are they really enough to account for the almost-surreal pace of diversification seen in the most extreme cases? Some models as well as intuition suggest there must be more going on. We have seen evidence from relatively small-scale studies that hybridization is a plausible candidate through the additional variation it can provide. The challenge, a daunting one, has been to identify unique, testable predictions about what patterns should be seen if hybridization and introgression have indeed played important roles in accelerating major adaptive radiations in ancient lakes. The largest genomic data sets currently available, with the sort of data essential for addressing such questions, are from the cichlids of the African Great Lakes.

A first step is to ask if there is evidence for hybridization at the start of a radiation to seed it, and in addition, ask if hybridization was ongoing as diversification proceeded. Lake Victoria, the youngest of these lakes, has almost certainly experienced the highest sustained pace of speciation and thus is the greatest challenge to explain. As more and better genome sequences accumulate, a process that is itself accelerating, research findings will surely grow more definitive, but important results have already appeared.

Substantial evidence does indeed indicate that hybridization between long-separated lineages of cichlids occurred during the early stages of the adaptive radiation in Victoria and associated lakes. The cichlids of these lakes, which include Lakes Edward, Kivu, and Albert along with Victoria and a few smaller bodies of water, are often referred to as the Lake Victoria Region

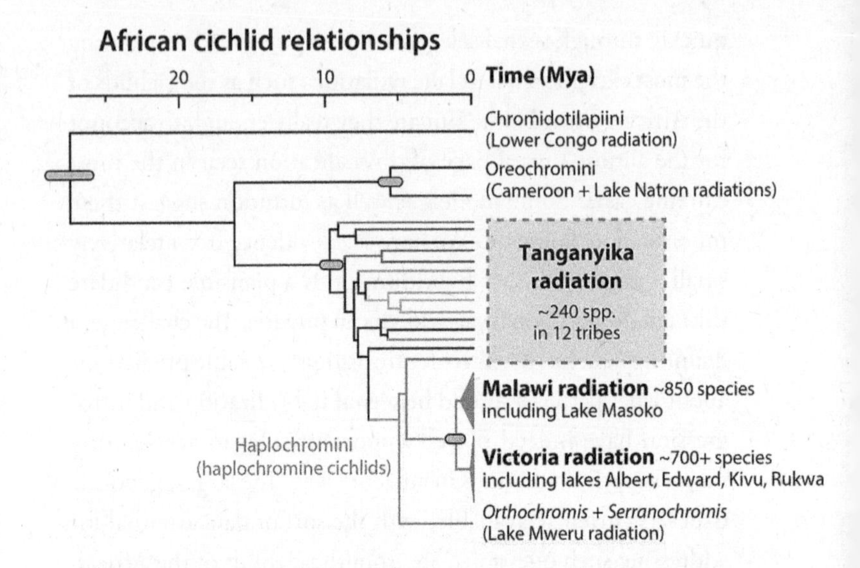

Figure 6.2

The relationships among African cichlids (hybridizations not illustrated), with particular reference to the African rift valley lakes. *Source:* Reprinted with minor modifications with permission from Springer Nature, Ronco et al., "Drivers and Dynamics of a Massive Adaptive Radiation in Cichlid Fishes," *Nature* (2020) as modified by Svardal et al., "Genetic Variation and Hybridization in Evolutionary Radiations of Cichlid Fishes," *Annual Review of Animal Biosciences* (2021).

Superflock (figure 6.2). One of the first hints of hybridization at the start of this radiation was the inconsistency between the results from analyses of mitochondrial DNA and genes from nuclear DNA, specifically for Lake Victoria species. The mitochondrial data suggest an extremely young and genetically homogeneous species group, whereas some of the nuclear data reveal more variation and hence an older diversification. This pattern could be explained if hybridization and introgression had occurred early on.

More extensive analyses of nuclear sequences sampled from across the genome of a wider range of African cichlid species have provided further support for the hybridization-at-the-start hypothesis. Joana Meier, who led this study with a group of collaborators from Bern, found that the nearest relatives of the species flock came from the Congo drainage, but not all genes told the same story. About 20 percent of the genome suggests a closer relationship to an upper Nile lineage. Their interpretation of this pattern is that substantial hybridization occurred between Congo and Nile lineages at the start of the radiation—a result in part of the active geology of the region and changes in connections between rivers. Critically, the patterns in the DNA are different from what would be seen if separate Congo and Nile lineages simply diversified alongside one another with a hybridization now and then. Rather, the analyses by Meier and her colleagues reveal that genes of Congo and Nile origin are often interspersed throughout a single individual's genome. Further, the proportion of Nile-origin genes is about the same across all the lakes, indicating that hybridization occurred in the common ancestor of the superflock and the introgressed genes persisted as the radiation proceeded. Because there was substantial evolutionary divergence between the Congo and Nile lineages prior to hybridization, the probability of transgressive variation in the hybrids is considerable.

One of the best-studied individual genes in the system, for the long wavelength-sensitive opsin, appears to have one major set of alleles with a Nile origin and another with a Congo origin. Differences between the opsins similar to those between *P. pundamilia* and *P. nyererei* pop up time and again throughout the Lake Victoria radiation. Generally, the shorter wavelength-sensitive form of the opsin gene is originally associated with a Congo lineage, whereas the longer (redder) wavelength-tuned allele has a mainly Nile history. In Lake Victoria, the longer wavelength-sensitive allele is generally found in species living in more turbid and/or deeper waters, just as in *Pundamilia*.

Matt McGee, Seehausen, and a large group of international and Bern collaborators, including Meier and others mentioned previously, have looked further into hybridization and diversification in Victoria and beyond, assembling and analyzing several large cichlid data sets in a landmark study. We have already reviewed some of their findings from an evolutionary tree that included sequence data from every species of cichlid with regard to ecological opportunity and sexual selection. They also conducted innovative analyses of 100 complete genomes sampled strategically from the more than 500 cichlids native to Lake Victoria and ecologically divergent pairs of species native to a set of lakes of varying ages that contain cichlid radiations.

One of their main efforts focused on genome evolution in Lake Victoria species for which conventional evolutionary trees could not be considered a given. Rather, these analyses assumed a network of species exchanging alleles with each other. If hybridization is frequent enough, this makes sense, since different parts of the genome will not have descended from the same ancestor, as in the branching pattern of evolution that is usually assumed. Each genome will instead be a mishmash of genes that came together at different times, some perhaps early in the radiation and others quite late, through a recent hybridization event. In these analyses, species with similar diets and habitats still shared more genetic material than expected by chance. Yet the results for stretches of DNA associated with color patterns and presumably carrying genes for color suggested that closely related species more often differ for color genes. This is consistent with the hypothesis that sexual selection and color divergence have played an important role in cichlid speciation in Lake Victoria.

McGee and colleagues also used analytic methods that enabled them to extract more information from the genome sequences than has often been the case in studies based on genome comparisons. They were able to more fully consider not just the sequences of the nucleotides that comprise DNA but large-scale insertions and deletions of DNA sequences too. Insertions and deletions are important and informative because they are usually a distinctive, even unique type of mutation, and large ones are infrequent. In general, the presence of the same major insertion or deletion in two species argues that either each inherited it from a common ancestor or there was a recent exchange of genes between the species, such as through hybridization. Because they are so distinctive, insertions and deletions can act like tags with which we can follow a chunk of DNA as it passes through a radiation, much as we might follow a radio-tagged bear across a landscape. In both cases, it is not the tag that is informative but rather what it is attached to.

Unlike a radio tag, an insertion or deletion will not always be attached to something of interest, but with large samples of such markers, some inevitably will be, and their distinctiveness will allow their histories to be traced. McGee and his collaborators treated insertions and deletions as indicators of relatively old genetic variation since they originate infrequently and should accumulate slowly. Sets of species with a great deal of such variation separating them are presumed to have been diverging for a long time or have obtained their ancient, divergent alleles in some other way, such as hybridization with other species. Applying this reasoning, McGee and colleagues calculated how many large insertions and deletions were present in the genomes of the ecologically divergent species they analyzed within each lake relative to the time separating the species sampled from that lake's radiation. They then compared their measure of insertion and deletion "enrichment" to a separate calculation of the rate of speciation in each lake. In the lakes with the most rapidly evolving radiations, especially Victoria and Malawi, there was more evidence of old genetic markers separating species than in the lake lineages with less rapid diversification, resulting in a positive correlation overall. The sample size of lakes is small, and avoiding spurious results can be challenging in such analyses, but it will be fascinating to see if the positive correlation they observed persists as data sets expand and the analyses are refined.

McGee and colleagues next concentrated on insertion and deletion variation within the 100 sequenced species from Lake Victoria. They emphasized old variants—the ones that also differed between lineages thought to be representative of the ancestors of the Victoria radiation. These markers were often correlated with particular ecological traits within the Victoria species, which would be expected if some of the individual insertions and deletions were located close to, or even within, genes that were adapted to particular ecological functions in the ancestral species. Presumably they were utilized in similar ways in Victoria cichlids, facilitating rapid adaptation and diversification. One particularly noteworthy association involved a region on chromosome 9 harboring a small set of markers. This region was "fixed"—that is, the only form present—in all fourteen of the Lake Victoria cichlids (in this study) that ate fish, an unusual diet for cichlids. It was also fixed in a predatory cichlid from Lake Kivu and a more distantly related lineage of predatory cichlids from southern Africa, *Serranochromis*. Given the relationships among these species, the allele may have originated as long as ten million years ago, possibly in fish eaters in a lake that has since disappeared. Exactly what role it plays in fish eating, however, remains to be worked out.

The genomic results for Lake Victoria may help to resolve the puzzle of how a lake that is relatively young (though its basin is a little older) can be home to a cichlid radiation comparable to those in the more truly ancient lakes of Malawi and especially Tanganyika. It appears that much of the variation underlying the ecological diversity of Victoria's cichlid radiation is in fact very old, and hybridization at the start of the radiation and subsequently has allowed it to be remixed into a wide variety of ecological types. The rapid speciation involved may have been facilitated by both ecological opportunity and strong sexual selection, with the latter linked to environmental variation as well.

Large-scale genomic analyses of the Lake Malawi cichlid radiations have also been conducted, led by Milan Malinsky, Hannes Svardal, and an international group of collaborators based at Cambridge University as well as several other institutions. Some Malawi lineages have diversified almost as fast as Lake Victoria's, and there are even more species in Lake Malawi than in the Victoria region superflock. The radiation is somewhat older too.

The Malawi radiation appears, again, to have involved hybridization at the start, in this case of lineages estimated to be separated by more than three million years of evolution, even more than for those founding the Lake Victoria radiation. The Malawi radiation is dominated by DNA closely related to that of a lineage that includes Lake Victoria superflock cichlids as well as a group of widely distributed riverine cichlids. The second lineage that contributed to the Malawi radiation is today represented by a river-dwelling species of the genus *Astatotilapia,* which has only an informal name, "Ruaha blue" (based partly on where it was collected). One of the more surprising observations that emerges from this work is that this hybridization could easily have been missed. The Ruaha blue lineage is known only from one species first collected in 2012, at a site a considerable distance from Lake Malawi. If the lineage had gone extinct or been overlooked, detecting and making sense of the hybridization would have been problematic. Thus, an immense quantity of genome sequences may be required to get a complete picture of hybridization history.

The Malawi data also suggest that basal hybridization, at the beginning of the radiation, was important to diversification within the lake. Svardal and colleagues evaluated genetic differences between species that live and forage along the bottom versus those that do so in the open water environment—a form of ecological divergence known to be ubiquitous in freshwater systems and crucial in the early stages of the Malawi radiation. They found that the genetic differences between the bottom-living and open water species disproportionately involved Victoria-associated DNA in one and Ruaha blue DNA in the other. They then looked throughout the genome at those genes for which a Ruaha blue version was present in some species and a Victoria variant in others. As for Lake Victoria, genes for opsins, which underlie color vision and are implicated in both adaptation to different lake environments and mating preferences, were particularly common among such genes.

The Malawi results also confirm the role of ongoing gene flow throughout the radiation. By comparing evolutionary trees suggested by different portions of the Malawi cichlid genomes, the research team inferred that after initial divergence into three main lineages, hybridization continued in Lake Malawi. Some hybridization was of intermediate age, not at the start of the radiation, but between the ancestors of multiple modern species. Recent hybridizations between sometimes distantly related species are suggested by their analyses too. Variation in lake levels and conditions may have contributed to intermittent periods of elevated hybridization, as different species and groups were brought together by declining water volumes and reduced water clarity increased interspecific mating.

I have so far said little about hybridization in the most ancient of the African Great Lakes, Tanganyika. Fabrizia Ronco and colleagues in Walter Salzburger's laboratory at the University of Basel have investigated this topic in the most comprehensive genomic analysis to date of the cichlids of an African Great Lake. They sampled 240 species, or nearly every species in the radiation. I saw her present the initial results of this study at a conference on the shores of Lake Victoria in 2018, and the thoroughness of the work was impressive and memorable. Ronco and colleagues observed considerable hybridization between species within major branches of the Tanganyika radiation, known as "tribes," though little between major branches. This is not so surprising given the great age of the major branches of the Tanganyika tree, but contradicts some earlier findings that suggested hybridization early in the Tanganyika radiation.

In a test of hybridization's potential role in mediating diversification rates among these branches, there was no clear effect. The speciation rate was not correlated with the hybridization rate. A different, potentially related pattern was present, however: the branches containing species with the highest average levels of heterozygosity—that is, genes with two different alleles—were the most rapidly speciating. Since hybridization would be expected to elevate heterozygosity and teasing apart these effects is difficult, a consequential role for hybridization in the radiation remains a possibility.

Considering the African Great Lakes together, there is considerable and growing evidence that hybridization at the beginning of the two younger radiations, in Victoria and Malawi, was important to their dazzlingly rapid diversification. Hybridization continued in each of the lake radiations and has likely made ongoing contributions to diversification.

Genomic data on the scale necessary to look reasonably comprehensively at the role of hybridizations in ancient lake diversifications only started to become available in the 2010s, and as I write, the African Great Lakes data sets are almost unique among large radiations. Still, the findings to date on the Telmatherinids of Sulawesi's Malili Lakes, ricefish speciation in the Malili Lakes and Poso (reviewed in chapter 2), and pupfish of Titicaca all support the concept of extensive hybridization in ancient lake adaptive radiations. I read almost daily about new initiatives to expand the breadth and depth of genome sequencing across the tree of life, so we can anticipate the proliferation of vastly larger data sets with which to explore more powerfully, extensively, and decisively the role of hybridization in adaptive radiation. In the meantime, a different, younger set of freshwater systems provide complementary data on the roles that hybridization and standing genetic variation can play in speciation and adaptation, and merit a short foray out of the oldest lakes.

THE TRANSPORTER HYPOTHESIS

Studies of smaller, younger bodies of water have advantages as evolutionary systems owing to the replication they sometimes provide and tractability of their smaller scales. Research on one inhabitant of such habitats, the threespine stickleback, is generating results that notably reflect and complement those emerging from ancient lakes. Stickleback lived in the drainage ditch in front of the house I grew up in, and I later did my PhD work with them, so I have a particular fondness for these prickly little creatures and their frantic, jerky courtship dances. Today they are probably the most extensively studied vertebrate in evolutionary biology even though, unlike the cichlids, they never form large species flocks.

What is special about the stickleback is that it has done certain things, including colonizing fresh water from the ocean, over and over and over again in thousands of different streams, rivers, and lakes on similarly extraordinary numbers of landmasses, from continents to tiny islands. In doing so, sticklebacks have evolved in parallel on a hemispheric scale, typically in response to lakes and streams opening up as glaciers retreated, but in a few instances finding newly available habitat when earthquakes moved coastal landscapes up and down. I have studied these charming fish in many locations, and one of my most memorable experiences was to see stream-resident stickleback in Gifu, Japan, for the first time. I knew that between British Columbia and Gifu, the stickleback of the Pacific Ocean were larger, bonier, longer spined, and more silvery than the dull little freshwater-resident fish scooting about the shallow streams of my hometown. Yet the Gifu fish, which my friend and collaborator Seiichi Mori showed me, could have been mistaken for the stickleback from the ditch in front of my parents' house. They were marvelous.

Considering freshwater sticklebacks from Japan, Europe, Alaska, and other distant locales, one might have guessed that the adaptations related to each freshwater colonization involved mainly different mutations, maybe even at different genes. A team led by Pam Colosimo in David Kingsley's lab at Stanford, collaborating with Dolph Schluter and his group, tested this possibility for one of the most conspicuous traits that often distinguish freshwater and marine sticklebacks: the bony plates along the side of the body. Marine stickleback have no scales, but possess prominent bony plates extending from just behind the head all the way to the tail. They look like the armored steeds of twelfth-century Europe, whereas the freshwater fish, especially in western North America, usually have just a few of these plates toward the head. Differences in predators and the relative cost of producing the plates probably contribute to selection favoring fewer plates in fresh water; the details are still debated somewhat, but what selection generally favors in each environment is clear. What is also clear is that plate genetics are unexpectedly simple.

In a series of experiments and comparative studies, Colosimo and her colleagues showed that a single gene, known as *Eda*, is responsible for much of the variation in bony plates. And to the surprise of many of us, the very same *Eda* allele reappears in freshwater lakes and streams all over the Northern Hemisphere, especially in Pacific-connected streams and lakes. The explanation is that the recessive low-plated alleles are at low frequency and thus almost always in a heterozygous state in the ocean, paired with a dominant full-plate allele that causes the heterozygotes to look much like every other fully plated marine fish. The current thinking is that the low-plate allele is advantageous in fresh water, but occasional hybridization with marine, specifically anadromous, stickleback results in hybrids and "leakage" of low-plate alleles back into the marine population. The result is that when marine stickleback colonize a new freshwater habitat, the low-plate allele, now favored, can increase in frequency and the low-plate form can reappear. Colonizations of fresh water, both experimental and accidental, have been watched in real time, and the low-plated form has been seen to reemerge from marine colonists, rising to almost 100 percent in just a few decades. This process, in which low-plated fish seem to pass invisibly from one freshwater site to another, reminded Schluter of how a person in *Star Trek*'s transporter room would dematerialize only to reappear in identical form in a new locale, leading to the irreverent name for this hypothesis (figure 6.3). It is now clear that many other freshwater genes also leak into fresh water and are reused, so *Eda* is easy to detect but not unusual.

Much like genomic regions associated with Lake Victoria's fish predation niche, which originally evolved for that role in a

Figure 6.3
The transporter hypothesis of threespine stickleback evolution. *Source:* James Vaughan, under a Creative Commons CC BY-NC-SA 2.0 license (https://creativecommons.org/licenses/by-nc-sa/2.0), with modifications by Dolph Schluter and the author.

different system and place, adaptation in stickleback frequently makes repeated use of the same genetic material in different freshwater systems. As a result, the low-plate *Eda* allele, like other freshwater-adapted alleles, is much older than the freshwater populations in which it often appears. Hence adaptation and even speciation do not always rely on new mutations but instead can make use of standing genetic variation, much of which has been around for a long time. It is interesting to speculate that the availability of more extensive, ecologically diverse ancient genetic variation may be a crucial difference between the tremendous breadth of Lake Victoria's radiations relative to the presence of just one or at most two species of sticklebacks in so many Northern Hemisphere lakes. Exploration of the planet's biodiversity at the genomic level has only just begun, and the data are accumulating at a breathtaking rate. More surprises are surely forthcoming.

* * *

We are accustomed to thinking of hybridization as an occasional problem for identifying animals in the field and a brake on biodiversity. But the jumbling of genetic material that occurs with hybridization may result in novel traits and lead to diversification along new ecological paths. Hybridization may have been an especially important yet long underappreciated factor in accelerating diversification in the most extraordinary ancient lake radiations. The reuse of ancient variation through hybridization may also help to resolve the enigma of Lake Victoria's cichlid radiation and how this young lake has come to show diversity more typical of truly ancient systems. Evolution may make use of standing genetic variation, some of it quite old, more frequently than was suspected and occasional exchanges of genes between species will facilitate such reuse.

7 SPLENDOR IN THE MUD

> *The existence of Lake Victoria's diverse endemic biota must be reconciled with the incontrovertible geophysical and paleoecological evidence . . . and not vice versa.*
>
> —J. Curt Stager and Thomas C. Johnson, 2008

I once gave a presentation on my Malili Lakes research in Bogor, Indonesia, to a group of university and museum scientists. I was delighted with how our work was going as we looked at the relationships between light environments, color patterns, and behavior to better understand how fish coloration evolved and diversity was maintained. So I was disappointed when a member of the audience asked how I could study evolution without looking at fossils and thus without direct information about how things had changed over time. I was taken aback and did my best to explain our approach, but it was a fair question. Most twenty-first-century evolutionary biologists are focused on the flora and fauna alive today. We mostly infer the history of organisms indirectly, estimating changes over time using DNA sequences, geographic distributions, and other features we can readily examine in the here and now. Lakes offer the possibility of looking at history more directly and ancient lakes offer extraordinary timelines.

The history available from lakes, relative to what can typically be gleaned from terrestrial records about life on land, is a bit like the information in a diary compared to that in a shredded letter. We can rarely find comprehensive terrestrial records of land-dwelling organisms. For a terrestrial creature to become a conventional fossil, dying in or near water helps a great deal. It then must be covered by sediments before it can be destroyed by other organisms or the elements. Of course, running this gauntlet is much more likely for bones and shells, which means that the fossil record is largely lacking for creatures without hard parts, like most worms, for example, or jellyfish in the marine environment. Next, the sediments containing the candidate fossil must be stable for the many thousands of years required for fossilization, at least in the geological sense. Finally, the rock containing the fossil has to be eroded just enough to come to the attention of a scientist, but not so thoroughly that the fossil is destroyed. There are oddities and exceptions, like frozen mammoths, tar pits, some caves, insects in amber, and a few others, but by and large this is the process. It involves a series of steps, each unlikely on its own, and the probability of the whole sequence is vanishingly low. One encouraging development for terrestrial systems comes from studies of ancient environmental DNA (free in the environment rather than inside organisms) from permafrost sediment samples. These have contributed to remarkable data sets on the distribution of plants and animals across the arctic over the last 50,000 years and provided important insights on a number of important topics, such as how long extinct species persisted.

Still, permafrost has a limited distribution, and samples from lake sediments are sometimes also included alongside the permafrost analyses.

In contrast, the sediments beneath an ancient lake will often comprise a layered, thorough record, older at the bottom to more recent toward the top, of what has been in the lake. The sediments accumulate gradually as material slowly rains down from above. If the lake is deep and stable, the water close to the bottom may also lack oxygen, which means few organisms other than bacteria will be present and decay will be slow. Whether oxygen is present or not, organisms that have died in or on the sediment, or drifted down from above, frequently just stay where they fell, as if waiting for a researcher to dig them up. I first encountered an ancient lake sediment sample while visiting my friend and colleague Doug Haffner at the University of Windsor in the early 2000s. Doug had returned from Indonesia with some of the first samples of Malili Lakes sediments and was incredibly excited about his treasures. As I came to understand the information that could be extracted from such samples, I better understood his enthusiasm.

While the information that can be gained from lake sediments is remarkable, the process of collecting the needed samples is daunting, involving amounts of money and logistical complexity well outside the comfort zone of many field biologists. These endeavors have as much in common with particle collider-scale physics, or large medical trials, as with the work Charles Elton, Eugene Odum, and the other architects of modern ecology did day-to-day.

At the heart of sample collection is a *corer*, a long tube driven vertically into the lake bed in order to retrieve a *core*, a

Figure 7.1
Drilling rig on Lake Ohrid. *Source:* Reprinted with minor modifications with permission from Elsevier, from Wilke et al., "Scientific Drilling Projects in Ancient Lakes: Integrating Geological and Biological Histories," *Global and Planetary Change* (2016).

long, hopefully intact sample of the lake bed sediments. The coring device is typically lowered from a barge (figure 7.1), which itself may be in hundreds of meters of water. Frequently the deeper parts of lake basins are the most stable and have the longest continuous records; thus, they are both informative and difficult to access. In the oldest lakes, sediments may extend many hundreds of meters beneath the lake bed, resulting in enormously long cores. Often parallel, overlapping cores must be collected to obtain a complete, "composite" core. The borehole itself is also of interest, and tools may be lowered into it to measure various physical properties of the sediments that surrounded the core. Not surprisingly, the machinery involved in this work is highly specialized. I once saw Thomas Wilke, who has played a lead role in addressing biological questions with the sediment data from North Macedonia and Albania's Lake Ohrid, give a presentation on his and his colleagues' drilling efforts there. They had to ship their massive, yet extremely difficult to replace, drilling equipment from Salt Lake City in the United States to an inland lake in the Balkans. This was a gargantuan logistical effort that inevitably had complications, including a fire onboard the transport vessel on the way to Europe.

Despite the many obstacles, several ancient lakes have now been drilled and had sediment cores collected, though for a long time a central goal in core collection and analysis was to obtain records of climate, and sometimes terrestrial vegetation. This resulted in much less information than later, more comprehensive efforts, although the work was a good deal simpler. In twenty-first-century studies, ancient lake coring projects often involve dozens of scientists from a wide range of disciplines, including geology, climatology, hydrology, evolutionary biology, fisheries, aquatic ecology, terrestrial ecology, and even anthropology. The anthropological insights are perhaps the most surprising products of extracting old mud from deep under a lake. But climate and vegetation records from East African lakes are providing increasingly detailed information about the conditions in which our own species and our recent ancestors evolved—as well as effects that our ancestors, in turn, had on their environments.

For those of us interested in the lakes themselves, it is the improved, sometimes entirely new approaches to discovering the history of the lakes and their denizens that are of greatest interest. To generate such records, a key initial task is to estimate the likely age of each portion of the sediment core, ideally in such a way that core samples from different portions of a lake can be matched with each other. Volcanic events of known age can play a valuable role here as they will leave a distinctive layer of ash, referred to as tephra, that can be detected in different cores and even different lakes. Tephra deposits can provide firm chronological anchors owing to frequently abundant potassium-rich minerals in these deposits, suitable for absolute radiometric dating. A variety of other approaches and data are also utilized in the dating effort, including the orientation of the earth's magnetic field, as indicated by the alignment of magnetic particles in a core's layers. The magnetic field periodically changes its polarity, and the timing of these changes is usually known, facilitating dating. Radiocarbon dating may be used for organic materials such as terrestrial plant fragments and pollen. In younger portions of the core, annual layers may be visible, although earthquakes, volcanic activity, and other disturbances can result in the loss or distortion of portions of the record. Hence many core studies report only quite rough dating, but methods have improved over time. No matter the details, the analysis of cores remains a slow, labor-intensive undertaking that can continue for years after collection.

Once the initial steps are completed and reasonably reliable age estimates are available for the various sections of a core, the biological and environmental data can be plotted against time. Physical and chemical properties of the sediments and borehole can provide information from throughout the lake's history regarding temperatures, salt content, rainfall, erosion around the lake, and other variables. Lake levels can often be estimated too, especially if some cores come from areas of a lake that have actually dried at times. Such drying leaves a strong, direct signal in the record.

Cores from shallower sections of a lake will draw the keen interest of biologists because many fish, mollusks, and other creatures are found mainly in such areas. Their downside is that shallow cores can be incomplete due to occasional drying. Entire skeletons of creatures as large as fish are sometimes encountered, but rarely, and even if present, they may be damaged or lost during processing. Smaller hard parts, however, are often abundant, to the point that they can be analyzed quantitatively throughout the time period covered by the core. Remains can include shells of mollusks and exoskeletons of crustaceans and insects as well as the bones, scales, and teeth of fish. Some microorganisms also leave large numbers of fossils, enabling especially powerful quantitative analyses. Pollen from land plants continues to be a major focus in core analyses owing to what it can tell us directly and indirectly about terrestrial ecological communities and even climates. In addition, it is now clear that many microscopic organisms live in the sediments themselves, and these too are getting more attention. This "subsurface biosphere research" has revealed slow-growing microbes deep below the bottoms of lakes and even beneath the seafloor. The biosphere is much, much larger than long assumed.

Shells, teeth, and other hard parts have tremendous value, but more can be done with the remains of animals than to identify and count such items, even if that process is sped up by automation. It has been possible for some years to retrieve and investigate the molecular remains of lake organisms, and the information to be gained from such efforts grows more and more sophisticated. The data arising can include proteins or other biomarkers, which may identify the sorts of organisms that were present, or reveal important information about their activities or metabolisms. The most exciting prospect, though, is the recovery of ancient DNA. Ancient DNA has been explored extensively with extinct terrestrial animals, to the point where efforts are underway to bioengineer a woolly mammoth, or at least a first approximation of one, based on genetic information from frozen animals. Ancient DNA investigations have also helped reveal our own hybrid history, involving Neanderthals and Denisovans. Such DNA has usually been retrieved from intact tissues of various sorts, particularly when the intent was to compile a genome. But "environmental DNA," which is no longer associated with an organism or its tissues, can sometimes also be retrieved and studied, as in the aforementioned permafrost studies. For biologists today, this is simply another tool—but of course it is much more than that from a slightly longer perspective. To scoop fossil DNA from the mud deep in a lake (i.e., "sedimentary ancient DNA"), sequence it, then quickly identify it through reference to vast genetic archives available from a computer connected to other computers through the ether might have sounded quite *Star Trek* or *Harry Potter*-ish just decades ago. Yet here we are.

One challenge associated with studying living microbes, environmental DNA, and many biomarkers is that the samples must be pristine (figure 7.2). Unfortunately, core samples can easily be contaminated, particularly since they are being collected from the bottom of a deep lake in likely a remote area; high temperatures, ultraviolet radiation, and even the lake water itself are a threat to the samples. And the problems

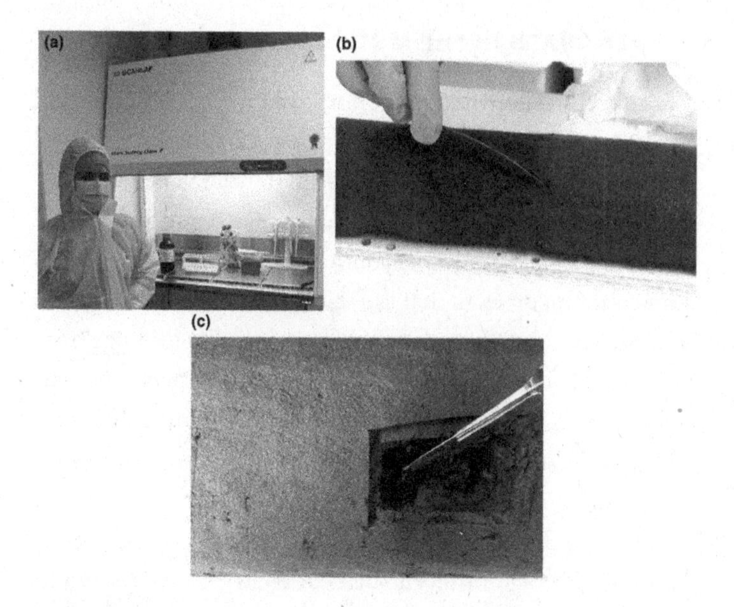

Figure 7.2

To avoid sample contamination, especially for analyses of ancient DNA, a researcher (a) wears a full body suit and other coverings before (b) scraping the potentially contaminated surface off a core and (c) collecting samples from its interior. *Source:* Reprinted with minor modifications with permission from John Wiley and Sons, from Parducci et al., "Ancient Plant DNA in Lake Sediments," *New Phytologist* (2017).

don't end with the acquisition of the core, even if it emerges from the collection process untainted. When DNA and other delicate molecules are to be analyzed, the core, or at least subsamples from it, will need to be kept under extremely cold conditions. All of this is a great deal easier if the field site is closer to well-equipped laboratories than is likely to be the case on many ancient lakes. Even so, these logistical challenges are being solved.

ALMANACS IN THE MUD

Fortunately, the tremendous effort involved in collecting and analyzing the sediment cores of ancient lakes has proven well worth the trouble. And the results are starting to come in.

One of the most important results yet to emerge from lake sediments is the astonishing youth of Lake Victoria. It has been known for many years that the start of the current interglacial period was a dry time in East Africa. Still, the findings from work led by Thomas Johnson of the University of Minnesota's Large Lakes Observatory were clear: Lake Victoria dried completely at least once as the polar glaciers retreated, possibly for a few thousand years.

In an influential publication that appeared in the journal *Science* in 1996, Johnson and his colleagues reported the results of an extensive set of cores from Lake Victoria. They complemented the core analyses with seismic reflection profiling, which involves sending acoustic waves through the water and lake bed, and then collecting and analyzing the reflected energy to map the subsurface; the principles involved are much like those of radar or sonar. The bottom line is that at several sites around the deepest parts of the lake, ordinary sedimentation dates back to about 15,000 years ago (the estimated timing has shifted a little over the years) and then there is a hiatus. Through coring, it was determined to correspond to a layer of terrestrial soil; there are even vertical plant rootlets growing through the deposits! Other indicators tell the same story, and among geologists it seems beyond dispute that the lake was dry for a time. Johnson and colleagues make the case that not only were the sites desiccated but it is also unlikely that any remnant lakes persisted within or along the boundaries of the modern lake, at least on a scale that might have supported ecologically diverse populations of cichlid fishes. Their reasoning is that because evaporation from Victoria itself contributed so much to atmospheric water and precipitation, there could be no lakes of any consequence in that vicinity without a substantial version of Victoria.

There was a good deal of discussion of Johnson and colleagues' conclusions for a time, some of it skeptical, but that seems to have faded and a broad consensus has become established that a major drying did occur. Any small water bodies that remained would have been more pond-like and salty, quite different from a large lake. Such clear evidence for the desiccation of Victoria, for some or all of the period from about 17,000 to 15,000 years ago, dramatically changed how biologists thought about the evolution of Victoria's cichlids. Mainly, it increased still further the estimates of the speed at which that radiation occurred.

Lake Malawi is older than Lake Victoria and its fluctuations tell a quite different story, but one that changes our take on evolution in that lake—the lake that houses more species of endemic fish than any other. The time calibration of Malawi cores has been notably extensive, reaching deep into its history. Sarah Ivory, then at Brown University and now at Pennsylvania State, worked on this project with a team of collaborators from multiple institutions, including the University of Arizona's Andy Cohen, one of the most influential, interdisciplinary scientists working on the analysis of ancient lake sediments (as well as the crab-snail coevolution discussed earlier). Ivory and her colleagues worked with a 380-meter (!) core, the length of about four US football fields, which itself came from beneath water 590 meters deep—which is almost half again as deep as the deepest point in North America's Lake Superior. Their calculations and analyses reveal a strikingly dynamic, almost-volatile lake history with features that illustrate processes also observed in other lakes.

The record worked out by Ivory and colleagues extends back about 1.2 million years, spanning the entire "modern" history of the lake. Although the Lake Malawi basin had the features of a deep lake beginning over 4 million years ago, it was largely dry from about 1.6 million years ago until the period covered by Ivory and colleagues' study. The early parts of the core suggest a relatively shallow lake, which alternated with even shallower marshy conditions, on a geologically short cycle of up to about 12,000 years. There was consistent riverlike flow through the system during both lake and marsh episodes. The main outlet, through which water left the lake, was the Ruhuhu River, which itself was connected with drainages that carried the outflow to the Indian Ocean on the east side of the African continent.

Starting about 800,000 years ago, the lake's cycles began to change, with riverlike characteristics and marsh stages disappearing. Instead, the deep phase of the lake involved deeper, blue water conditions with a strong, stable layering of the lake's waters. During such phases of stratification, the depths of the lake lacked oxygen. These conditions, much like those of the modern lake, alternated over longer periods of 20,000 years with a shallower lake that was saltier, more alkaline, and exhibited thorough mixing rather than stratification. It also had higher algae levels and therefore is referred to as the "green lake" phase.

What happened to trigger these changes and a different set of cycles? One key development likely started with a geological event. Ivory and colleagues suggest that tectonic uplift, a movement in the surface of the earth, lifted the edge of the lake basin above the point of access to the Ruhuhu River (figure 7.3). This resulted in a new, higher outlet at the Shire River, which is the current outlet, and a higher lake level. It is as if a bucket had two holes in its side, one low and one high. If the low one gets plugged, the bucket can hold more water before it starts to leak. Thus, during some portions of the climate cycle, which had entered a phase in which periods with high lake levels were longer, the lake filled with water to the new outlet and took essentially its modern form. It became highly stratified, with low oxygen in its depths and extensive blue water. In addition, its shallow, rocky outcrops were covered by water, resulting in the complex underwater geography of the modern lake. This is biologically important because it has been suggested that these rocky patches, separated by areas of sand and mud, enhance isolation, speciation, and diversification in some rock-loving cichlids.

When there was not enough water entering the lake, however, its level could not get high enough to access this new, elevated outlet. Thus, during drier portions of the climate cycle, there was insufficient water in Lake Malawi for the new outlet to be reached. With no flow through, salts became more concentrated in the lake. At this time the lake was also well mixed all the way into its depths, ensuring that key nutrients were available for algae and leading to green, turbid water. It was like some disturbed modern lakes, where water is diverted for irrigation before reaching the lake, even as nutrients pour

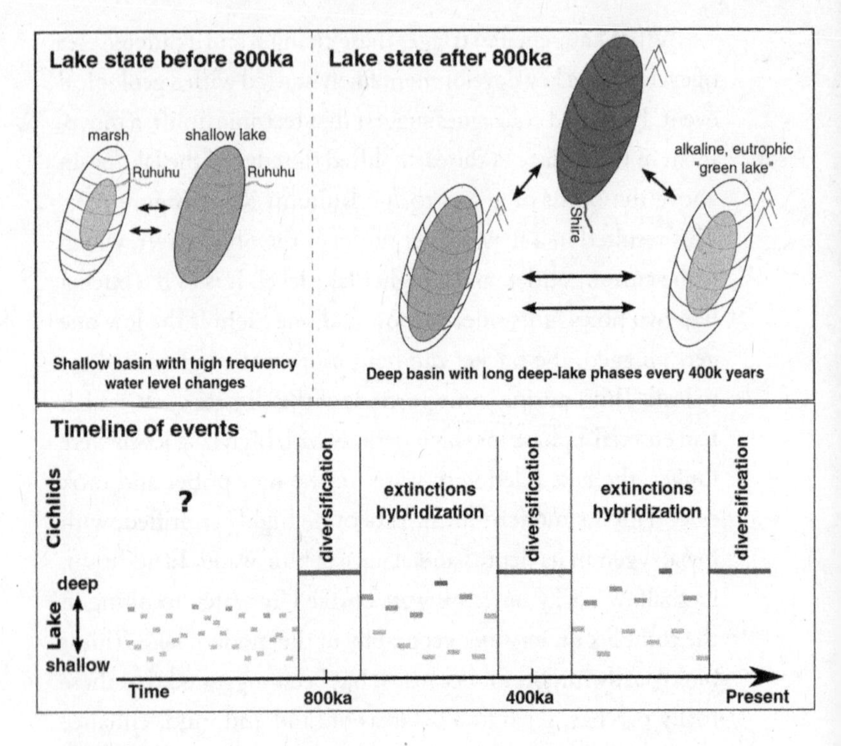

Figure 7.3

Lake Malawi. The vertical axis in the lower panel shows lake depth, with cichlid evolution characterized above each period. Before the benchmark 800,000 years ago ("ka" is thousand years), shallow-deep cycles were shorter and the lake was always shallower than it would be later, especially during the extended deep lake periods at 800,000 years ago, 400,000 years, and recently. The dark-colored lake in the right upper panel is deep and clear. Between the three deep lake periods, lake levels cycled more rapidly with saltier, more alkaline, high algae conditions when the levels were low (illustrated by the contracted, lighter-colored lake). *Source:* Reprinted with minor modifications with permission from the National Academy of Sciences of the USA, © 2016 National Academy of Sciences, from Malinsky and Salzburger, "Environmental Context for Understanding the Iconic Adaptive Radiation of Cichlid Fishes in Lake Malawi," *Proceedings of the National Academy of Sciences* (2016).

in from agriculture or dwellings with inadequate sewage treatment. Ivory and colleagues suggest that these low–water level, saline, turbid phases were likely periods of increased extinction and hybridization among the cichlids whereas the blue water phases would have been times of diversification (figure 7.3).

The blue water phases were all longer than they were prior to 800,000 years ago, but they were not all of the same duration. Exceptionally long blue water episodes have occurred on a 400,000-year cycle, with the first starting around 800,000 years ago when the modern lake was being established, and the last currently underway. Ivory and her colleagues make the case that it was only during these extended blue water phases that extraordinary diversifications such as the modern one could unfold. Overall, the scenario they have presented is broadly consistent, and satisfyingly so, with the major genomic analyses reviewed in the previous chapter. The initial radiation began around the time that the modern lake appeared about 800,000 years ago or a little later, followed by repeated episodes of hybridization and introgression.

There is one other finding from the Malawi studies that bears special emphasis: the change in the outlet river that occurred with the transition to the modern deep lake, or at least the modern deep-lake cycle. Whereas previously the water exiting the lake had flowed from the north end almost directly east toward the Indian Ocean, after the uplift the outlet shifted toward the south and the Zambezi River system. The outflow still eventually ended up in the Indian Ocean, but along a quite different routing via the Zambezi drainage. Such changes in connections between water bodies and drainages occur in many freshwater systems, and can result in the mixing of faunas, with hybridization and introgression, or isolation and divergence of species and populations. Changes in drainage connections are often profoundly significant biological events.

Some of the features documented in the Lake Malawi studies have reappeared in analyses of cores from Lake Towuti, the largest and most biodiverse of Indonesia's Malili Lakes. These Towuti cores were collected in 2015 by another large international consortium, led by James Russell of Brown University and Hendrik Vogel of the University of Bern. Some initial results appeared quickly, but it was several more years before many of the major analyses and findings began to be reported, illustrating the time-consuming processing and analysis involved with long cores. One of the key results to emerge was an estimate of the age of the lake, which previously was based on only rough calculations. Russell, Vogel, and their colleagues, including Haffner and Thomas von Rintelen, whom we met earlier, estimate the age of the lake to be about 1 million years, comparable to the modern Lake Malawi.

Modern Lake Towuti exhibits extremely low biological productivity, but the cores show that this has not always been the case. Today, algal growth appears to be limited in part by unusually low levels of phosphorus, an essential nutrient. Phosphorus is, for example, a principal ingredient of most garden fertilizer and key component of the agricultural runoff that sometimes results in pea-soupy lakes overwhelmed by algal blooms. One cause of Towuti's low phosphorus is the unusual, metal-rich water chemistry of the Malili Lakes. In both Towuti and Matano, high concentrations of iron oxides interact with phosphorus to make the latter unavailable to diatoms. Much of the phosphorus ends up in the sediments and essentially removed from the system. The permanent stratification of the modern lake, distinct from temperate lakes, which typically experience seasonal mixing, further ensures that deep-lake phosphorus never becomes available to diatoms and plants in shallower water where photosynthesis can take place. Diatoms, which are important in Towuti and other ancient lakes, are single-celled algae that possess robust silica shells that preserve well in lake sediments. Diatoms are mainly bottom dwelling, but are sometimes found in open water.

Towuti's "green" periods, at least two substantial ones over the last million years, occurred when diatom abundance in the water column increased to high levels. The sediment core layer containing one of these periods, evocatively described as "diatomaceous ooze" in the publications, sits on top of the tephra from a volcanic explosion, whereas the other such layer is close to tephras but not adjacent. Russell and colleagues interpret this as evidence that phosphorus in the tephras, which may have eroded slowly and released material for an extended period, likely contributed to the bursts of diatom abundance. Thus, volcanoes may have played a key role in the ecology of the lake. However, changes in how phosphorus cycled within the lake and its availability to diatoms were likely also important, and may have resulted from complex interactions of physical and biological processes. More detailed analyses of the species composition of the diatoms by Mariam Ageli and colleagues, based mainly at Canada's University of Windsor, have emphasized such nuance, including the potential role of variation in seasonal patterns of water column mixing.

As with Malawi, these results from Towuti's sediments are broadly consistent with what we know about the main fish radiation there, involving the Telmatherinidae, or sailfin silversides. The current estimated age for Towuti is in the general ballpark for the time frame of the telmatherinid radiations there, but to say something more definitive, it will be important to learn more about the other lakes in the system, some of which may be older, and further investigate telmatherinid divergence times. What is unequivocally striking is that once again there are periods of murky and ecologically disrupted conditions in a system with a history of hybridizations. Unlike the African cichlids, color is not currently known to be significant in maintaining barriers between species in the telmatherinids, but changes to water transparency and chemistry could also affect mating patterns based on traits other than color, such as shape and behavior. Water chemistry changes might even influence the perception of chemical cues known to be crucial to courtship and mating in many fish.

Thinking back to my own experience of Towuti, one of the features that today distinguishes it from Lake Matano, where fish biologists have more often worked, is the presence in Towuti of saltwater crocodiles. Saltwater crocodiles are infamous for hunting and eating people, including in nearby Australia. In Indonesia, they seem to be less feared and in some places are treated with reverence. Unfortunately, crocodiles do sometimes attack people in Towuti. My friend and collaborator Fadly Tantu lost a fisherman friend to a Towuti crocodile. This was especially sad because it was probably preventable; he had been spearfishing at night. Hence Towuti is more problematic for making extensive underwater observations of fish and other organisms, though some undaunted scientists have nonetheless spent quite a bit of time in the lake (and really, the crocodiles tend to be localized). In coming years, underwater drones with video cameras could potentially make the entire lake more readily accessible for collecting observational data on fish, so as to complement the extensive sediment record and the further results that will undoubtedly emerge from it.

BIBLES IN THE MUD

The Thompsons of Manitoba, my mother's family, had a family bible in which births, deaths, marriages, and other key events were recorded with dates. This practice is a common one. Today, evolutionary trees inferred from living species can give us a rough equivalent of the Thompson bible's records for the lineages presently found in ancient lakes. These trees, however, tend to miss the evolutionary equivalents of the deaths in the family bible, which are extinctions. Trees constructed using DNA data from living species can be pretty fuzzy on dates too. Consequently, a quantitative record from the lake sediments of what species lived there and when, regardless of whether or not they are alive today, could be an extraordinary step forward for evolutionary biology. Further, it is a step almost unimaginable with many research systems.

Important results of this sort are already emerging from Lake Victoria's sediments. Moritz Muschick, working at both the University of Bern and Swiss Federal Institute of Aquatic Science and Technology, and an international team of collaborators including Russell of the Towuti cores, Johnson of the Victoria desiccation work, and Ole Seehausen, have studied fish remains, especially teeth (figure 7.4), from the earliest postdrying period of Lake Victoria. They reviewed samples

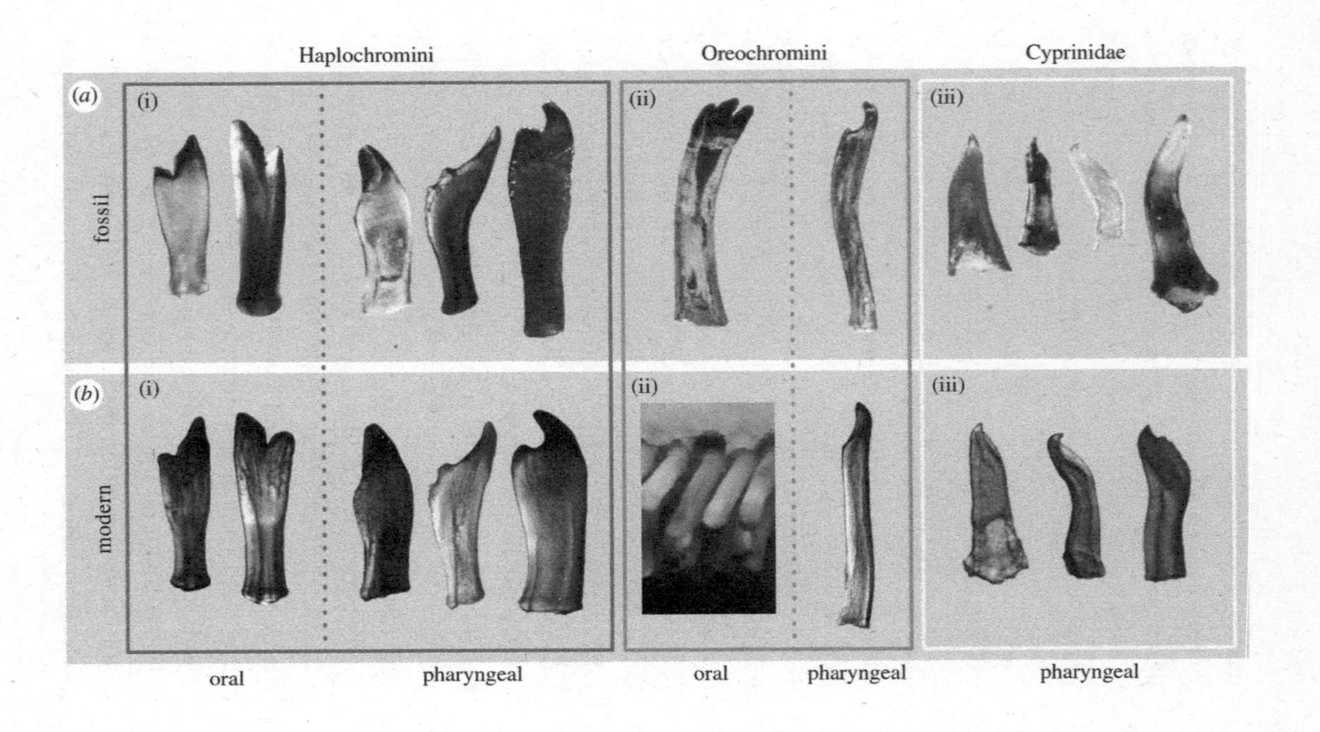
Haplochromini
Oreochromini
Cyprinidae
(a)
fossil
(b)
modern
(i)
(ii)
(iii)
oral
pharyngeal
oral
pharyngeal
pharyngeal

from some of the same 1994–1996 cores collected by Johnson and colleagues for their influential study of the lake. The cores had been stored in a long-term facility in Duluth, Minnesota. Long-term retention and preservation, a now-standard practice, helps ensure that the maximum possible information is garnered from valuable and difficult-to-obtain core samples. There can be a downside to such extensive use, though, as Muschick and colleagues report fungal growth and concerns about carbon contamination. Fortunately, the fish teeth that the team were studying were unlikely to be affected.

The ability to identify fish teeth from the cores enabled Muschick and colleagues to address a long-standing question about ancient lake radiations. Was the success of the fastest-radiating lineages, here haplochromine cichlids, simply a result of their priority? Of being the first group in an empty habitat and adaptively radiating into available niches before other groups arrived? Often it is hard to know exactly which groups first became established during lake formation, but with cores

Figure 7.4

Fossil teeth from (a) Lake Victoria sediments and (b) modern counterparts. Haplochromini are the cichlids that massively radiated in the lake, while Oreochromini are cichlids that did not radiate. The Cyprinidae (did not radiate) teeth are from an open water plankton feeder. Oral teeth are the familiar teeth readily visible within the fish's mouth whereas pharyngeal teeth come from the pharyngeal jaws, which are located in the throat. Images are not to scale. *Source:* Reprinted with minor modifications with permission from the Royal Society, and with permission conveyed through Copyright Clearance Center, Inc., from Muschick et al., "Arrival Order and Release from Competition Does Not Explain Why Haplochromine Cichlids Radiated in Lake Victoria," *Proceedings of the Royal Society B: Biological Sciences* (2018).

and the ability to identify teeth it should be possible to confidently address this question.

Muschick and colleagues' main result was surprising: the haplochromine cichlids indeed appeared just after the lake started filling, but so did other fish lineages. They found that two potentially effective fish competitors appeared in the lake at the same time as the haplochromine cichlids—the group that would undergo the fastest sustained adaptive radiation yet known among vertebrates. The competitor lineages were oreochromine cichlids, commonly known as tilapia, and cyprinids, a family that includes minnows and carps. Both groups have undergone radiations in other lakes so they have at least some potential for diversification. Moreover, the species of tilapia, cyprinids, and also catfish that were present in the early lake occupied particular feeding and habitat niches into which haplochromines subsequently diversified. It is possible, and arguably likely, that there were other consequential ecological differences between the evolving haplochromines and their apparent competitors, despite dietary overlap. In any case, the other groups remained and were successful, but did not exclude the haplochromines.

This finding does not refute the ecological opportunity hypothesis of adaptive radiation by itself, but it does show that the haplochromines did not occupy a lake entirely empty of fish competitors and that simple opportunity does not adequately explain the success of the haplochromines. Certainly it raises the possibility that the capacity of the haplochromines to evolve and diversify quickly has been critical to their success. One feature sometimes suggested to be important to this ability in cichlids, including haplochromines, is illustrated in figure 7.4: the pharyngeal jaws and teeth, a second set of jaws and teeth present in the throats of ray-finned fishes (the vast majority of fishes). In 1973, Karel Liem hypothesized that modifications in cichlids to their pharyngeal jaws, making them more powerful and versatile, are a "key innovation" that facilitated the independent evolution of the familiar oral jaws relative to the pharyngeal jaws, thereby facilitating ecological specialization and diversification. Subsequent studies have continued to emphasize the importance of their pharyngeal jaws to the ecological versatility of cichlids, but there may be greater integration between the oral and pharyngeal jaws than in Liem's original vision.

The next task in Lake Victoria research is to trace the diversification of the haplochromines through the history of the lake. Nare Ngoepe, a PhD student in Bern, is leading an analysis of the fish fossils, working with Muschick and Seehausen on more recently collected cores. Her study of teeth identified using their visible features is being complemented by investigations based on DNA extractions and analyses. In other younger lake systems and the ocean, environmental DNA from sediments has begun to yield results for timeframes shorter than those of most ancient lakes. In two Swedish lakes, sediment DNA showed different histories of postglacial colonization by whitefish over a scale of about 10,000 years, while in Beppu Bay, Japan, variation in the abundance of anchovy, sardine and jack mackerel over the last 300 years was broadly similar in datasets derived from sedimentary DNA and more traditional methods. Approaches using ancient DNA will likely prove informative in ancient lakes too; the question will be how far back in time reliable results can be obtained, and whether their quality will be sufficient to allow analyses to go beyond identifications that rely on individual genes, for example to look at patterns of hybridization. The technical challenges are substantial, but the insights generated could be extraordinary. The Lake Victoria researchers have presented some initial genetic results, and other studies have made use of environmental DNA to study Tanganyika fish in the modern lake.

In Lake Ohrid, which straddles Albania and North Macedonia, studies of diatoms in sediment samples have enabled some of the most comprehensive analyses of diversification and extinction yet reported for an ancient lake radiation, or perhaps any radiation. Diatoms are exquisitely complex in form—miniatures of abstract art. Some electron microscope images from Ohrid's sediments are shown below (figure 7.5), but I encourage the diatom-smitten reader to also look at the many color images available with a few keystrokes and an internet connection.

The Ohrid record is exceptionally long at 1.36 million years and extends from the formation of the lake to the present. Wilke worked with a large international team to sample diatoms across the entire 447-meter (of the composite core) Ohrid record, at a resolution of 2,000–4,000 years (fig. 7.6). They encountered and quantified 152 species endemic to Ohrid, or about 75 percent of those known to have occurred in the lake, tracking the appearance and disappearance of each as measures of speciation and extinction.

In the lake's early history, when it was still widening and deepening from 1.36 to 1.15 million years ago, both the

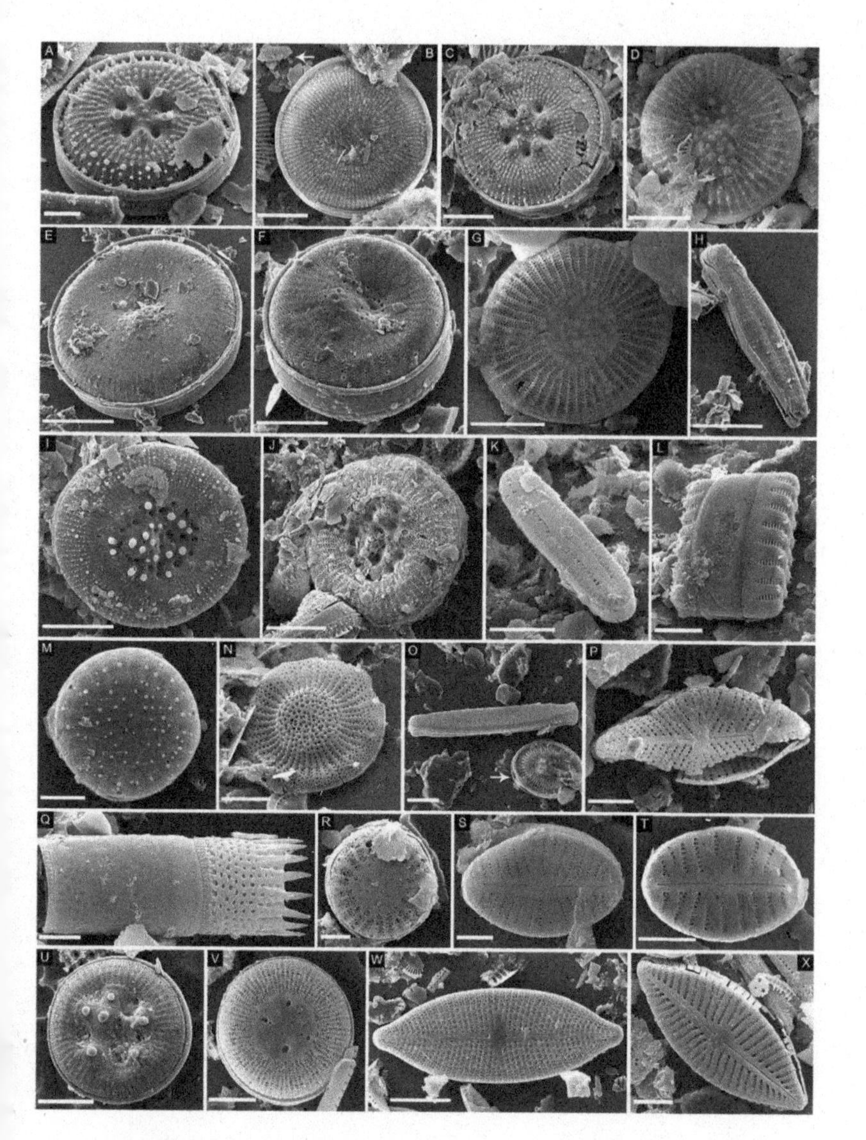

Figure 7.5

The fossil diatom diversity of Lake Ohrid (scanning electron microscope image, scale bars 1–10 μm). *Source:* Reprinted with minor modifications with permission from AAAS, from Wilke et al., "Deep Drilling Reveals Massive Shifts in Evolutionary Dynamics after Formation of Ancient Ecosystem," *Science Advances* (2020), © the authors, with some rights reserved; exclusive licensee AAAS. Distributed under a CC BY-NC 4.0 license (http://creativecommons .org/licenses/by-nc/4.0).

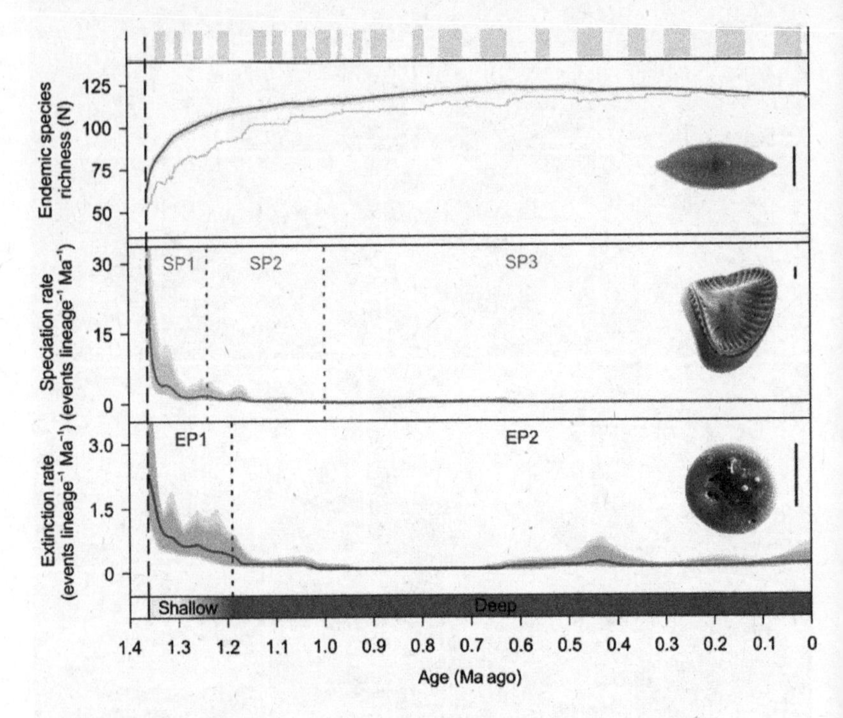

Figure 7.6

Changes in extinction rate, speciation, and species richness for endemic Lake Ohrid diatoms over 1.36 million years. The darker bars at the top of the figure indicate periods of glaciation; the light bars indicate interglacial periods. Representative diatoms are pictured. *Source:* Reprinted with minor modifications with permission from AAAS, from Wilke et al., "Deep Drilling Reveals Massive Shifts in Evolutionary Dynamics after Formation of Ancient Ecosystem," *Science Advances* (2020), © the authors, with some rights reserved; exclusive licensee AAAS. Distributed under a CC BY-NC 4.0 license (http://creativecommons.org/licenses/by-nc/4.0).

formation and extinction of species occurred at high rates, as shown in the lower two panels of figure 7.6. The net effect was for species richness to increase, as shown in the top panel. This increase continued, even as both the speciation and extinction rates diminished, eventually plateauing. The change in extinction rates is perhaps the most novel part of this study and was not obviously predicted by theory—although it makes intuitive sense when conditions are changing. Wilke and colleagues interpret the gradual plateauing of overall diversity alongside the decline in the speciation rate as resulting from the diatom community approaching the lake's ecological limits to species diversity. It is striking that these processes were so steady and consistent even as glaciers toward the poles came (the dark bars at top) and went (the light bars), and the authors stress the buffering effect of the deep lake.

These findings suggest considerable stability, at least as the lake is experienced by its endemic diatoms. It will be fascinating to compare long-term sediment fossil data sets from other ancient lakes to Ohrid's as well as patterns for different groups of organisms.

* * *

We have mainly studied evolution using observations and samples from modern organisms, or by assembling the best data we could from the often-sparse records in sedimentary rocks. Ancient lake sediments allow us to directly assess nearly complete histories of aquatic environments and even organisms over long timeframes. Sediment data show that Lake Victoria's cichlid flock evolved in an astonishingly short time and provide an environmental context for repeated episodes of hybridization in Malawi. Fossil fish teeth have shown nascent Victoria to have been less ecologically vacant than anticipated by theory. Ohrid's diatom records enable the incorporation of extinction as well as origination into models of biodiversity, where extinction results differed from expectation. Sediment DNA results are just starting to arrive, but will likely soon form a cascade. They could be revolutionary.

8 THE BLUE EYE OF SIBERIA

От Байкала начинается сибирская поэзия.

Siberian poetry begins from Lake Baikal.

—Anton Chekhov, in letter to A. N. Pleshcheyev, 1890

Baikal, or the Sacred Sea as it is known in Russia, is the largest, deepest, and most ancient of freshwater ancient lakes. At Baikal's beginning, usually considered to have been about 25 million years ago, our planet was a quite different place—the fauna, climate, and even continents were not as they are today.

If, for example, you were somehow to find yourself by the North American seaside of that time, you might encounter *Pelagornis sandersi*, a seabird with a wingspan of over twenty feet, about the same as an Andean condor and a wandering albatross placed wing tip to wing tip and measured together. Venture inland a little ways and wildlife would likely have been common, as human hunters would be far in the future, but the mammalian grazers would have seemed odd. Camels, or at least members of the camel family, were abundant. South America, where the Camelidae had not yet arrived and given rise to llamas, had an even more exotic fauna in the late Oligocene and early Miocene (the Oligocene-Miocene boundary was 23 million years back, just after Baikal appeared), but would have been hard to get to. The Isthmus of Panama was still beneath the waves, and the seas were far from welcoming. *Megalodon*, the fifteen meters or so relative of today's great white sharks, was widespread.

Still, South America might well have been worth the trip. Among the stranger sights would have been a collection of gigantic, plant-eating, armadillo-like creatures—the glyptodonts. Your greatest worry might have been a bird, two or three meters tall and flightless—a sort of ostrich meets *Tyrannosaurus rex*. Known colloquially as terror birds, various species of these dinosaur-esque hunters preyed and scavenged across much of South America all the way through Baikal's early years.

Primates were well established in both South America and the old world 25 million years ago, but our own family, the Homininae, would not appear on the drying savannas of Africa for another 15 million years or more. Even the apes (the Hominoidea) had only just appeared, splitting off from the other old-world monkeys and their kin.

The weather was balmy in the late Oligocene, and some of the higher latitude regions had much more pleasant winters than they have today. In general, the climate varied less between the poles and equator. Ice had by this point covered Antarctica in a permanent blanket, but the glaciers later to dominate the far north would not become established for many millions of years. While the continents were close to their current positions, they were not quite there. The Americas, for example, were a little closer to Europe and Africa than they are today;

the Atlantic was still opening up, a result of slow but steady spreading along the mid-Atlantic Ridge.

Baikal too was very different 25 million years ago. In fact, while the age of Baikal is typically given as 25–30 million years, the rift in which it is located first appeared and filled with water much earlier, about 70 million years back. Thus, an early form of Baikal, archeo-Baikal (figure 8.1), may have had dinosaurs along its banks, perhaps the eight-meters-long duck-billed dinosaur *Amurosaurus*, for example (figure 8.2). Archeo-Baikal possessed a moist, subtropical climate, and while probably ecologically significant in its time, we would not recognize it as the modern lake. It was a string of smaller bodies of water, at most tens of meters deep, not consistently connected and occupying only the southern basin of today's lake. Still, there is some evidence that a few of the invertebrate lineages found in Baikal today got their start in the time of archeo-Baikal.

It is the Baikal that took shape 25–30 million years ago that is widely treated as recognizably today's lake, despite differences from the Baikal we know, so that is the time frame I have emphasized when discussing the lake's beginnings. Sometimes known as proto-Baikal (figure 8.1), it was initially two separate lakes, one in the modern south and central basins, more or less, and one in the farther reaches of the north basin, with the intervening areas largely dry. Owing to uplifting of surrounding land, however, the lake had become deeper, reaching depths of hundreds of meters in both sections.

During the proto-Baikal period, the climate remained much warmer than ours, but it was nonetheless drying and cooling. Along the shores of the lake, forests were gradually replaced by steppe and desert-steppe landscapes. The lake itself

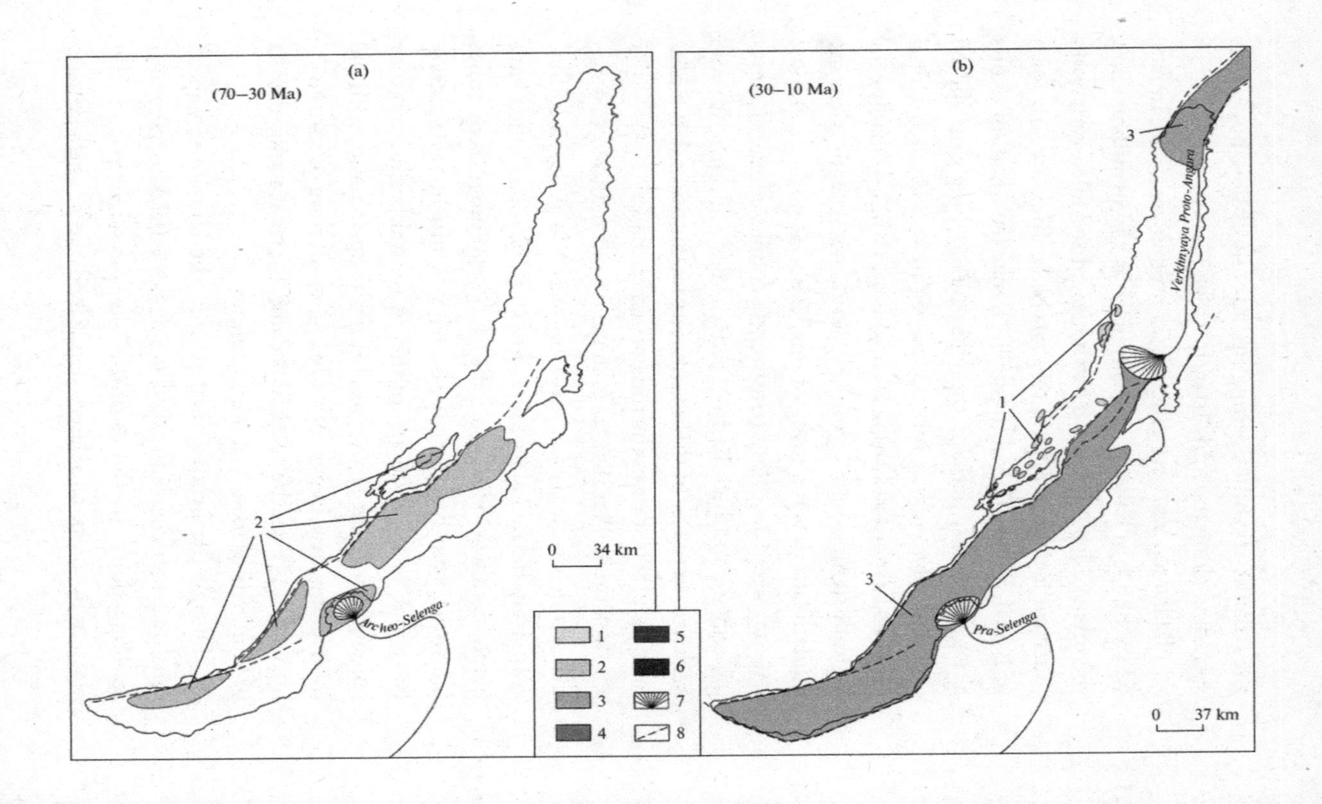
(a)
(70–30 Ma)
2
Archeo-Selenga
0 34 km
1
2
3
4
5
6
7
8
(b)
(30–10 Ma)
3
Verkhnyaya Proto-Angara
1
3
Pra-Selenga
0 37 km

Figure 8.1

The early days of Lake Baikal (the outline of the modern lake is shown, but only the shaded areas had water): (a) archeo-Baikal stage, close to 70 million years ago; (b) proto-Baikal stage, about 25 million years ago. Depths and substrates: (1) a few meters deep, (2) up to tens of meters, (3) up to a few hundred meters, (4) up to several hundred meters, (5) up to 1,000 meters, (6) > greater than 1,000 meters, (7) deltas, and (8) main faults. *Source:* Reprinted with minor modifications with permission from Springer Nature, from Mats et al., "Late Cretaceous-Cenozoic History of the Lake Baikal Depression and Formation of Its Unique Biodiversity," *Stratigraphy and Geological Correlation* (2011).

was not static during the long period of proto-Baikal and continued to evolve. Approximately 10 million years ago it became a single, continuous body of water with the submergence of the entire northern basin. Other parts of the lake continued to deepen, reaching over 500 meters.

Dramatic changes occurred in the climate and lake starting about 3.5 million years ago, which also marks the end of the

Figure 8.2

Amurosaurus riabinini, a duck-billed dinosaur found in eastern Russia at the time of archeo-Baikal. *Source:* Institut Royal des Sciences Naturelles de Belgique.

proto-Baikal period. With continued cooling and the onset of cycles of "ice ages" about 2.6 million years ago, Siberia moved toward its modern climate, or an even colder one during episodes of glaciation. Baikal, though, was never glaciated, even when regularly covered by ice during both glacial and interglacial episodes, including the interglacial we are in today. It did become much less productive when the glaciers grew, with diatom abundance in sediments greatly reduced as ice a mile deep spread across the northern reaches of the planet. It also continued to deepen as the Pleistocene proceeded, reaching about1,000 meters in the south basin approximately 1.6 million years ago, at the latest. But its depth went up and down with the advances and retreats of the Pleistocene's massive blankets of ice too. The ice affected precipitation, winds, and the flow of rivers around Baikal and across most of the planet.

It was surprisingly recently, in only the last 150,000 years, that Baikal became the ultradeep, consistently oxygenated lake that we know today, now 1,642 meters at its deepest point. There are a few other cool, deep lakes in the northern reaches of our planet, some fairly old (see figure P.1), but only Baikal has evolved a distinctive deepwater fauna and an open water ecosystem dominated by a unique set of organisms.

OFFSHORE BAIKAL

Some of the key ecological players in Baikal's extensive offshore waters are utterly unlike those of any other water body. And they seem to get odder as you work your way up the food chain.

The base of the food chain, at least in terms of animals, is a species endemic to Baikal but not especially peculiar: the

1.5-millimeter copepod crustacean *Epischurella baikalensis* (known until recently as *Epischura*). Copepods are tiny crustaceans (the same group as crabs, shrimp, and crayfish) that are usually abundant in large lakes and seas, though most lakes lack an endemic species. When I think of these creatures, I always think of North Atlantic right whales because a former lab mate of mine spent many hours poring through copepod samples in order to better understand the ecology of right whales, which were feeding on the copepods. I still find it perplexing that despite their diminutive dimensions, these tiny crustaceans can be abundant enough to support warm-blooded animals as massive as whales. In Baikal, there are of course no whales (there is a seal, which we will come to), but *Epischurella* is indeed ecologically dominant among zooplankton, the lake's tiny open water animals, and enormously important as food for larger creatures. It is not common in sheltered bays, especially during the summer, but in the open waters of the lake, *Epischurella* often comprises more than 90 percent of the zooplankton. It feeds on a mixture of open water algae and microorganisms such as ciliates (with ciliates possibly more important than long assumed), and is popularly credited with keeping the water clean and clear.

The great majority of the *Epischurella* are found, year-round, in the upper 250 meters of the water column—a range extending deeper than the bottom of most lakes, but of course encompassing only the shallower portions of Baikal. In the summer, *Epischurella* exhibit "diel vertical migration," coming within about 5 meters of the water surface at night and heading deeper during the day. This is a widespread phenomenon that is observed in the zooplankton of many water bodies. It is generally thought of as an adaptation to avoid visually hunting predators during the day, when the waters just below the surface are well lit, while allowing feeding on shallow-dwelling algae at night. *Epischurella* is well adapted to Baikal's chilly waters and will tolerate only a narrow range of temperatures, doing poorly above about 15°C.

There are exceptional long-term data sets for *Epischurella* as well as other planktonic species and the waters they inhabit owing to a remarkable long-term sampling effort. Three generations of Siberian biologists, all members of a single family, began collecting physical and biological data in Baikal in 1945 and continued into the twenty-first century (figure 8.4). At least once a month, and usually every seven to ten days, they collected nine sets of samples at preestablished depth intervals from 0–250 meters at a site about 2.7 kilometers offshore, where the water is 800 meters deep. Their collection site is a little way off the head of the Angara River, which is shown in figure 8.3, upstream from Irkutsk. This would be an impressively consistent record in an easily accessed subtropical location—but in a sometimes frigid and difficult locale like Baikal it is heroic. Winter sampling required walking over the ice at temperatures that would be daunting for even my hardiest Canadian relatives. Occasionally the ice was not thick enough to walk on yet was impassable by boat as well, resulting in virtually the only situations in which sampling did not take place. Fortunately, such conditions were infrequent and temporary.

Data collection persisted not just through difficult weather but also through the reign of Soviet leader Joseph Stalin, the end of the Soviet Union, and a collection of other national and international upheavals and transitions. The initiator of

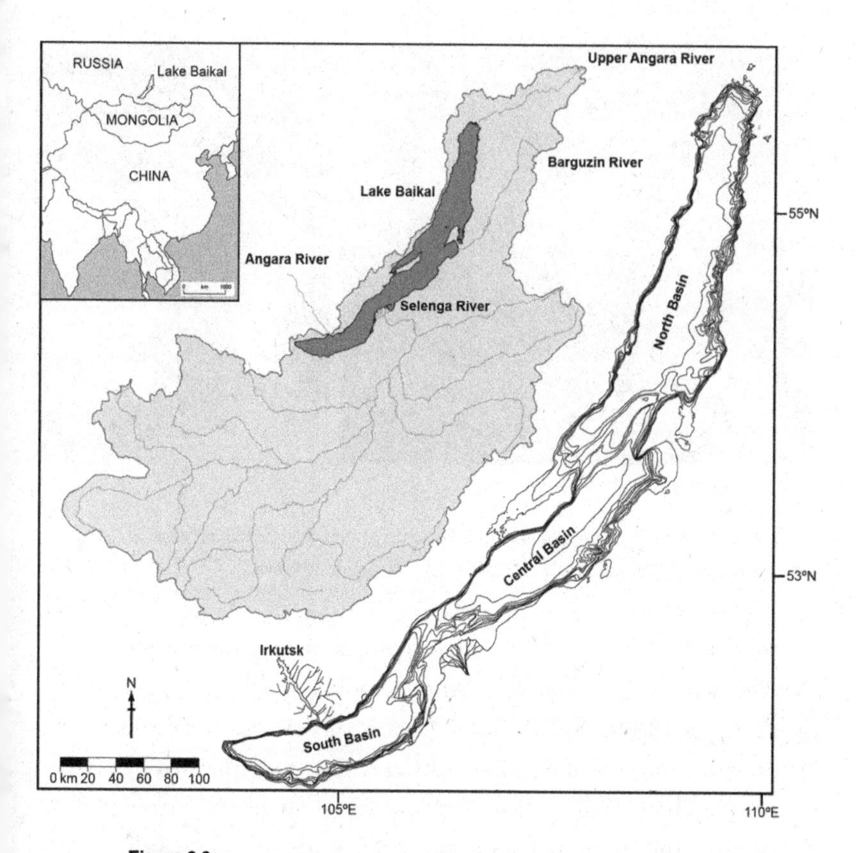

Figure 8.3

Modern Lake Baikal and its major drainages, with depth contours. *Source:* Reprinted with minor modifications from Swann et al., "Changing Nutrient Cycling in Lake Baikal, the World's Oldest Lake," *Proceedings of the National Academy of Sciences* (2020). Published under a Creative Commons CC BY 4.0 license (https://creativecommons.org/licenses/by/4.0).

Figure 8.4
Left to right: Lyubov Izmest'eva, as a child, with her grandfather, Mikhail Kozhov, 1955; Olga Kozhova, about 1975. *Source:* Lyubov Izmest'eva.

the effort was Mikhail Kozhov, a professor at Irkutsk State University and author of a famous book about Baikal from the early 1960s. His daughter Olga Kozhova continued the program and took a position at Irkutsk State too, followed by her daughter, Lyubov Izmest'eva, who has now retired from that institution (though data collection continues at a slightly reduced frequency). The family's long-term data have enabled a variety of analyses, including quantitative assessments of how zooplankton communities have changed with a warming climate and shrinking ice cover. Some of this work suggests a decline in *Epischurella* relative to other zooplankton species as the lake has warmed, which we will return to.

The next link in the food chain is also a crustacean and member of the zooplankton, but larger and more unusual. This is the gammarid amphipod, *Macrohectopus branickii*, a creature up to about thirty-eight millimeters long and a peculiar outlier in one of Baikal's most extraordinary radiations. Among the more than 265 species of gammarids in Baikal (the number is tough to pin down), smaller radiations in Titicaca and a few other lakes, and another radiation among the caves and underground waters of Europe, this is the only freshwater gammarid amphipod to have taken up an entirely open water existence and become a key component of an offshore food chain.

The evolution of *Macrohectopus* and the other Baikal gammarids was touched on in the first chapter, but the entire group merits a little more attention. If nothing else, the gammarids are noteworthy for being a truly ancient-ancient lake radiation. Exactly how long Baikal's gammarids have been in the lake is not known with accuracy, as the dating of their radiations is a work in progress, but they may be as old as the modern lake itself and could have been in archeo-Baikal.

The idea that Baikal's gammarid radiations were the result of repeated colonizations by non-Baikalian lineages has long been accepted, but there has been little agreement about just how many times this happened. A trend has emerged, however, for estimates to settle at two colonization events. This includes the report of the most comprehensive study yet, based on extensive sequence data, by a large international group including Sergey Naumenko at Moscow State University and Lev Yampolsky at East Tennessee State University as well as a follow-up study also involving Yampolsky. Their work emphasizes the radiation that resulted from the second invasion, the Acanthogammaridae and allies. This contains much of the ecological diversity seen in Baikal's gammarids, such as the "abyssal" species found in the depths of the lake and species

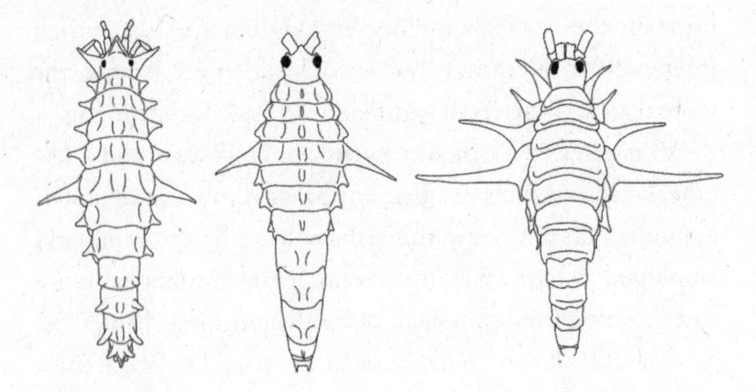

Figure 8.5

Left to right: Convergent body armature in gammarids from the Caspian Sea (*Axelboeckia spinosa*), Lake Baikal (*Acanthogammarus lappaceus*), and Lake Titicaca (*Hyalella armata*). *Source:* Reprinted with minor modifications with permission from Springer Nature, from Copilaș-Ciocianu and Sidorov, "Taxonomic, Ecological and Morphological Diversity of Ponto-Caspian Gammaridean Amphipods: A Review," *Organisms Diversity and Evolution* (2022).

that live parasitically inside the brood chambers of a different amphipod (brood chambers are a little bit like the pouches of kangaroos). It also includes those gammarids most associated with Baikal: heavily armored species like *Acanthogammarus victorii*, which are convergent with species in Titicaca and the Caspian Sea (figure 8.5).

One finding emerging from studies of this main Baikalian gammarid radiation is that speciation and diversification rates were highest close to its start. This is satisfying in its consistency with theory, which predicts most rapid diversification when the ecological opportunity is greatest—sometimes known as an "early burst." But it makes settling on the relationships among the major lineages, which diverged then, especially difficult. It also creates challenges for working out the relationships with non-Baikalian gammarids. Things happened fast at the start followed by millions of years of further evolution that tend to throw a genetic smoke screen in front of the earliest events. It is as if evolution were playing the children's game of telephone, where one child whispers a phrase to another one, who then whispers it to another child, and so on. After passing between a dozen or so players, the phrase may be changed yet easily recognizable or almost entirely different. The more players in telephone, or the more generations since the burst of an adaptive radiation, the greater the potential for change and the more difficult it is to make out the phrase, or relationships, at the start.

The older but less extensive radiation, known as the Micruropodidae, contains mainly species that are "fossorial," living buried in the sediments most of the time and possessing smooth bodies well suited to such a lifestyle. Unexpectedly, it is in this group that repeated analyses have placed the open water–living *Macrohectopus*. It is a genuinely puzzling result that a species with a wildly distinctive form and ecology should be found to have evolved from a group in which shape, body features, and ecology are relatively homogeneous and based on a life within the sediments—rather than from a larger set of species exhibiting much greater diversity. There is no clear explanation for this odd state of affairs as yet, but one observation may prove telling. One of the burrowing species, *Micruropus wahli*, has been caught in large quantities in surface waters at night. Perhaps habits of this sort in members of the Micruropodidae facilitated the evolution of a free-swimming open water feeder. It is likely also pertinent that in this older lineage, there was abundant time for *Macrohectopus* to diverge and adapt to its new lifestyle.

In addition, over such a long expanse of time, related species that might have clarified the evolution of *Macrohectopus* may have gone extinct in one of the many ecological upheavals that punctuate Baikal's long history. Analyses of sequence data indicate strong natural selection has taken place on the genes of *Macrohectopus*, possibly in support of adaptations to an active planktonic mode of life as well as the low temperatures and high-pressure characteristic of the deep open waters in which it often occurs.

Macrohectopus is so abundant in the modern Baikal that it is frequently compared to krill, the crustacean of the Southern Ocean that is the key food source for a long list of marine mammal and bird species as well as the focus of a human fishery. Abundant and large enough to have substantial energetic value, *Macrohectopus* is a critical food for several Baikal fishes, including the omul, the whitefish that has been the basis for Baikal's most important fishery (and I understand why; omul really is delicious).

Like the smaller plankton on which it feeds, *Macrohectopus* performs vertical migrations. Occupying a wide range of depths from the surface to 700 meters, it typically moves from deeper waters during the day to the upper 50 meters of the water column at night. The density of its aggregations is noteworthy, as is the synchronization of its vertical migrations, which can involve ascents as rapid as 4 meters per minute. Keeping in mind this pace corresponds to one to two hundred body lengths, it is about the same as a human swimmer of 1.5 meters height rising at 150–300 meters per minute. Anyone who has tried to free dive to any depth or even just do some speedy lengths in a pool can appreciate just how fast that is.

Not every *Macrohectopus* makes exactly the same migration, though. It is the large females that move the farthest. Males, which are only a fraction of the size of females, tend to make a shorter vertical migration. The movements of the juveniles, who typically do not go as deep during the day, are also less extreme than those of the adult females.

One of the most abundant fish in Baikal feeds extensively on *Macrohectopus* and is equally exceptional, as it is also an open water species from a group that almost always sticks to the bottom. It is also simply a strange creature, the golomyanka, *Comephorus baicalensis*. It is the most unusual member of a radiation of sculpins, an ancestrally marine group of fishes that have evolved into 100 or so freshwater species across the Northern Hemisphere, but that are nowhere as diverse in ecology or form as in Baikal.

Most freshwater sculpins, members of the superfamily Cottoidea, live along the bottom of streams, rivers, and lakes in relatively shallow waters. In Baikal, they have diversified into about thirty-three species found nowhere else, evolving in two novel ecological directions: into the depths, the abyss of the lake, and the open water zone. Given how thoroughly they dominate the fish community of each zone and how few competing species are present, the ecological excursions of this lineage of Siberian sculpins are likely another example of opportunity-driven adaptive radiation. The sculpins comprise well over half of the fifty-two species and subspecies of fish that are native to the lake.

Compared to the great age of the gammarid diversification, the radiation of the sculpins is relatively recent. Estimates of the radiation's timing are still being refined, but it seems clear that it took place in the essentially modern version of Baikal, most likely between roughly one and five million years ago, but a few million years earlier is quite possible. In another departure from the patterns seen in the gammarids, Baikal's sculpins appear to have all evolved from a single ancestor. Thus, they are all more related to each other than to any non-Baikal sculpin. Most likely they evolved from another species of freshwater sculpin from within, more or less, the genus *Cottus*. This placement suggests that the various Baikalian sculpins, currently assigned to different families, will need to have their classification revisited. One point of overlap between Baikal's sculpins and gammarids, though, is the tempo of their diversification. The sculpins diversified rapidly at the start too. Moreover, these groups are linked ecologically in that different species of gammarids are the main prey items for the various species of sculpins; their radiations are connected.

One of the reasons an open water sculpin is considered such an exotic creature is that among fishes, sculpins are notably ill-suited for this way of life. An open water sculpin is almost as biologically incongruous as a celery-feeding shark. The main shortcoming of sculpins when it comes to open water life is that they lack a swim bladder, the organ that most fishes living off the bottom use to maintain buoyancy. Through fine adjustments, many fishes dial up or down the gas present in this organ in order to stay neutral and level, neither sinking nor rising, much as a scuba diver does with a buoyancy compensator. Too little gas in the diver's inflatable vest and the diver sinks; too much and they rocket to the surface, which can be even more dangerous. Sculpins cannot make any such adjustments so they have had to find other ways to keep a stable position in the water column without expending undue amounts of scarce energy. They especially need to avoid sinking, the natural tendency of fish adapted to living on the bottom (still, partially open water sculpins less extreme than Baikal's occur in at least two North American lakes, though seldom mentioned in the literature).

There are a few species of open water sculpins in Baikal, but it is the golomyanka (the big one; there is a second smaller species that we will come to shortly) that shows the most extreme adaptations for maintaining neutral buoyancy. Whereas bottom-dwelling sculpins can have bodies containing less than 3 percent body fat, the adult golomyanka is commonly over 40 percent fat by volume. This is a remarkable amount; a fatty hamburger, for example, is about 20 percent fat, or 30 percent at the most. Cheddar cheese, one of the fattiest of cheeses, is about 33 percent lipid. Brie, which seems to me awfully rich, is about 28 percent. They are all much lighter fare than golomyanka. The golomyanka also has bones that are less mineralized than those of bottom-dwelling sculpins. Reduced bone mineralization and thus density is likely even more important for buoyancy in the other open water sculpin species, none of which have lipid levels at all close to those of the golomyanka. In addition to its high fat levels and reduced bone density, the golomyanka is similar to some deepwater marine fish in having almost no pigmentation. Therefore under some conditions, the tail end of a golomyanka can be eerily translucent.

In most sculpins, females lay their eggs underneath stones; males spray their sperm over the eggs in order to fertilize them and then defend them as they develop. The larvae may be bottom or open water dwelling, depending on the species, but they eventually develop into bottom-dwelling adults. But the golomyanka and the closely related *Comephorus dybowskii*, the little golomyanka, have also modified their reproduction for a fully open water existence. Instead of the male releasing sperm over the eggs after they leave the female's body, the two species have evolved internal fertilization and live birth, completely emancipating both of them from any link to the bottom of the lake.

Both species also show a predictable dietary progression as they develop. At small sizes, they mainly eat the copepod, *Epischurella*, and then shift to the open water gammarid *Macrohectopus* as they grow. This remains the overwhelmingly dominant prey item for little golomyankas, but as they grow larger both species eat golomyanka larvae as well as the larvae of other pelagic sculpins. For the little golomyanka, fish larvae seem to be an infrequent food, but for the larger species they are sometimes the main prey item. Both species participate in the daily vertical migrations of Baikal's open water fauna, following their copepod and gammarid prey as they move toward the surface at night and the depths during the day. The golomyanka go deeper than some, however—sometimes reaching the abyssal depths of the lake. While doing research on coral reef fish, I have sometimes hung in the water of a passage leading from the open ocean to the lagoon of an atoll as schools of open water fishes move onto or off the reef. It is an awesome sight, watching all of that life move past. I imagine that if one could hang in Baikal's water column, it would be an equally beautiful and much eerier procession as the translucent golomyanka make their way up or down, along with dense layers of *Epischurella* and *Macrohectopus*, at twilight and dawn.

Unfortunately for those living alongside Baikal, the golomyanka are solitary creatures and thus difficult to fish profitably. They are of great value, though, to the next step up the food chain to another odd creature to find in a freshwater lake.

Baikal's most famous denizen is the nerpa, *Pusa sibirica*, the only species of seal confined entirely to fresh water. There are seals that occur in freshwater lakes in other places—such as eastern Canada and northern Europe—but they are clearly populations of familiar marine species that occasionally colonize short-lived postglacial lakes. There is one other lake, the Caspian Sea, that is home to a species of seal, but a seal in a vast salty lake usually labelled a "sea" seems less noteworthy than Baikal's nerpa. Anyway, the Caspian seal, *Pusa caspica*, is closely related to the nerpa.

So how did a seal come to inhabit a lake over a1,000 kilometers from the nearest ocean and even more distant if one follows the rivers? The most likely scenario is that the seals swam there from the frigid seas along the north coast of Siberia via the Yenisei River and then Angara River, into which Baikal drains. The DNA sequence data, and shared parasites too, suggest that Baikal seals evolved from adventurous ringed seals that managed to make this ambitious trip. The Caspian seal likely had a similar origin, though separate and independent from the nerpa's. Sequence analyses suggest the nerpa lineage broke off from the ringed seal well after the radiations of the gammarids and sculpins, about 1.15 million years ago, give or take 500,000 years. Thus, the nerpa arrived in Baikal decidedly after the start of the Pleistocene and its cycles of glacial advance and retreat. River flows varied greatly with changes in the glaciers, and huge temporary lakes sometimes came and went. It is possible that these dynamics played a role in the colonization of Baikal by ringed seals, and the Caspian too, although no well-supported scenario has yet emerged.

Owing to Baikal's size and climate, nerpa have been able to establish a life cycle and ecology surprisingly similar to that of other ice-loving seals. These relatively small seals, which reach about 1.65 meters and 130 kilograms, live mainly in the northern and central basins of Baikal, most of the year staying in open water well away from the coasts. Adult animals winter alone on the ice using the thick claws on their front flippers to maintain a hole through which to access air when they are in the water. In February or March, females bear young in a lair made among the snowdrifts and nurse them for 1.5 to 2 months—longer than ringed seals. The life span of the nerpa is also relatively long, sometimes over 50 years. These extended suckling times and life spans are likely a reflection, at least in part, of a great advantage to a seal of life in Baikal: a paucity of predators. Ringed seals, the closest relative of the nerpa, are a key food item for several large predators of far northern seas, including the polar bear and killer whale; Greenland sharks also eat ringed seals. In departing the Arctic Ocean, the nerpa left these voracious and effective predators behind. There are brown bears at Baikal, and they may take an occasional hauled-out seal, but their hunting is nothing like that of the polar bear, an essentially marine animal that achieves much of its annual caloric intake while hunting from the sea ice. Humans hunt nerpa and have likely done so for a long time, but the population seems so far to tolerate the numbers taken, some of which are hunted legally and others illegally.

The resilience of the nerpa to all manner of human insult may well be the seal's most remarkable characteristic—resilience not just to hunting but also to the effects of viruses that we and our animals may have helped spread, and pollutants such as dioxin and mercury. Amazingly for a large mammal in such a constrained environment, the nerpa is an International Union for Conservation of Nature "species of least concern." Some of the population's hardiness may be a result of a stable food source, as nerpa eat a great deal of golomyanka, both the big and small ones. Because there is no fishery for these solitary open water sculpins, their abundance has not been depressed by human removals, which is a serious problem for other marine mammals that compete with human fisheries. But the seals also eat the even more abundant *Macrohectopus* amphipods, and a detailed study of the seal's feeding behavior, published in 2020, has revealed that nerpa are better adapted to this food source than was long appreciated.

Yuuki Y. Watanabe of Japan's National Institute of Polar Research, Eugene Baranov of the Baikal Seal Aquarium in Irkutsk, and Nobuyuki Miyazaki of the University of Tokyo temporarily attached biologging packages to several foraging nerpa in 2018. Each package contained equipment that recorded aspects of the seal's environment and swimming, including a video recorder. Marine mammals that eat open water invertebrate prey typically do so by catching many at a time using adaptations like the baleen of right whales, which sieve copepod prey by the thousands. The data from the foraging seals, however, revealed that they caught *Macrohectopus* one by one—but *fast*, at the highest rates yet recorded for single-prey-feeding aquatic mammals. Seals caught an average

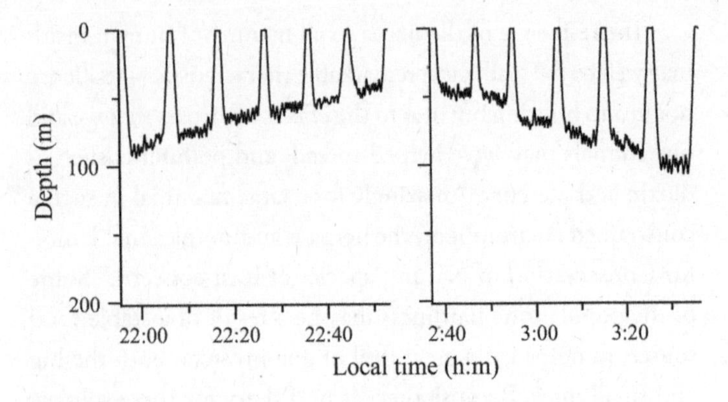

Figure 8.6

Depth data for the foraging dives of individual nerpa in June 2018. As the *Macrohectopus* ascend at dusk, the seals' move shallower, and their foraging depth decreases even within a single dive. The reverse happens as the *Macrohectopus* descend, with dawn's approach (note dusk comes late and dawn early at Baikal's latitude in the summer). *Source:* Reprinted with minor modifications with permission of the National Academy of Sciences of the USA, © 2020, National Academy of Sciences, from Watanabe et al., "Ultrahigh Foraging Rates of Baikal Seals Make Tiny Endemic Amphipods Profitable in Lake Baikal," *Proceedings of the National Academy of Sciences* (2020).

of fifty-seven *Macrohectopus* per dive, with each dive lasting about ten minutes. This led to the consumption of sometimes thousands of these gammarids per day.

Hunting was so efficient in part because the seals followed the vertical migrations of their prey, feeding on the gammarids mainly at night when they are shallower. In addition, they tracked the dense layer of migrating gammarids as it moved over the course of the evening and even during a single dive (figure 8.6). This highlights the pace and synchrony of these migrations.

Nerpa may also be aided in their high-speed snapping up of tiny prey by teeth that appear to be specialized for this purpose (figure 8.7). Like a few other seal species that feed on open water crustaceans, nerpa have comblike postcanine teeth. These likely allow the large amounts of water taken in during rapid foraging to be efficiently expelled while retaining prey.

The convergence with crabeater seals, which forage on krill in Antarctic waters, is especially striking, as is the contrast between the teeth of the nerpa and those of the closely related Caspian seal, which does not feed on amphipods. It remains to be seen if nerpa can suck in prey at a distance, as crabeater seals can. A captive crabeater was observed to suck in prey from half a meter!

Watanabe and colleagues suggest that the nerpa foraging on gammarid amphipods, rather than on the fish that eat the amphipods, has important ecosystem implications as well as for the nerpa. By eating lower on the food chain, the nerpa can take advantage of higher prey abundance owing to the energy that would have been lost in the transition from amphipods to fish. This may help to explain the exceptional abundance of nerpa in a deep, cold, unproductive lake; nerpa densities are several times higher per square kilometer than the densities of Caspian seals or two other freshwater seal populations. The nerpa's ability to survive and maintain high densities in a stressed, limited environment by eating lower on the food chain may carry a message for us too. Our odds of lasting a little longer in our own stressed, high-density habitat will likely be higher if we move down the food chain a link or two.

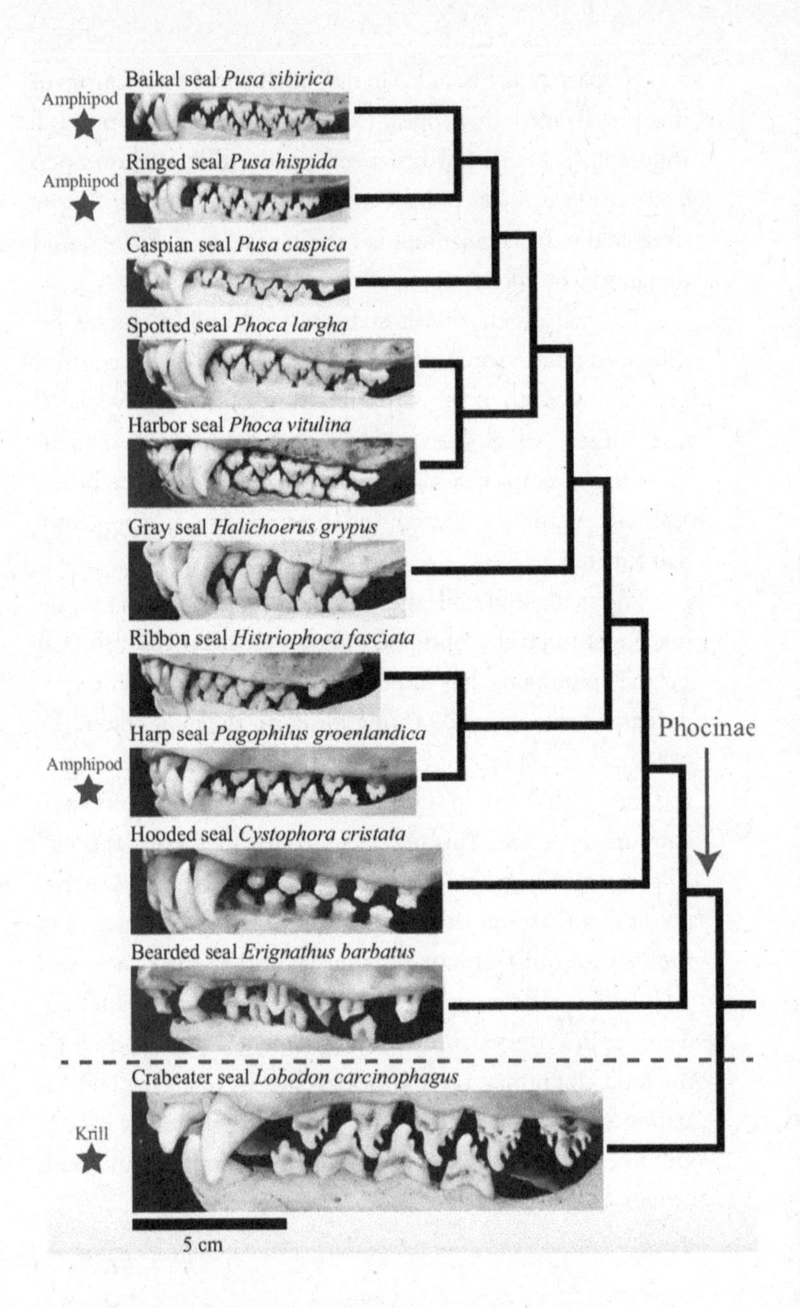
Baikal seal Pusa sibirica
Amphipod
Ringed seal Pusa hispida
Amphipod
Caspian seal Pusa caspica
Spotted seal Phoca largha
Harbor seal Phoca vitulina
Gray seal Halichoerus grypus
Ribbon seal Histriophoca fasciata
Harp seal Pagophilus groenlandica
Amphipod
Hooded seal Cystophora cristata
Bearded seal Erignathus barbatus
Phocinae
Crabeater seal Lobodon carcinophagus
Krill
5 cm

THE ABYSS: DESERT AND OASIS

Baikal's sponge forests are distinctive and beautiful, and its open water food chain is like no other. But among freshwater lakes, its most unique feature is its abyss.

Oxygen levels in the deepest areas of Baikal are often slightly lower than at the surface, but a great deal higher than in the depths of other ancient lakes and more than sufficient to support mature fish, amphipods, and other large invertebrates. Light from the surface, however, is almost absent below about 400 meters, meaning there is no photosynthesis and most food, which is sparse, must come from above. Temperatures are relatively constant, from about 3.2 to 4.0°C. Thus, the environment is stable, if difficult.

In the abyss, below about 400 meters, Baikal's fishes are all endemics and all sculpins. Other fish just do not go so deep, with pressure being one important reason. The majority of sculpin species occupy a wide depth range, including shallower waters that receive natural light and at least some of the deep, dark waters; about nine species are confined to shallower waters and do not venture into the abyss. About six bottom-dwelling

Figure 8.7

The jaws and teeth of the nerpa and other seals, with feeding on krill or amphipods indicated, and their relationships. Note how different the Nerpa's teeth are from those of the closely related Caspian seal. *Source:* Reprinted with minor modifications with permission of the National Academy of Sciences of the USA, © 2020, National Academy of Sciences, from Watanabe et al., "Ultrahigh Foraging Rates of Baikal Seals Make Tiny Endemic Amphipods Profitable in Lake Baikal," *Proceedings of the National Academy of Sciences* (2020).

species (i.e., excluding the golomyankas of open waters) live overwhelmingly below 400 meters, including two that are generally found below 900 meters. In their authoritative overview of Baikal's biodiversity, Kozhova and Izmest'eva provide an unappealing portrait of these deepwater sculpins: "[They have] a flabby body, covered with a tender skin easily gathering in folds . . . [and their] eyes are mostly reduced. The body is colorless or light pale-yellow and, as a rule, there are no spots." In the gammarids, the vast majority of species are either native to shallow waters or can live in a wide range of depths. The number of species found in the abyss is small, and the number found there exclusively is smaller still. Kozhova and Izmest'eva describe some common features of the abyssal species: "[The] eyes . . . lack pigmentation or are pale pink, but their antennae, as a rule, are very long. Body coloration is usually whitish or pinky-white."

Other Baikalian invertebrate radiations are less well studied than the gammarids, but are known to be represented in Baikal's dark depths. These include annelid worms, among them *Tubifex* sp., which are related to the worms sold as live food in aquarium shops, and planarian flatworms that include a set of frequently carnivorous deepwater species up to thirty centimeters in length. For those of us who know planarians as tiny creatures sometimes used in biology teaching laboratories, usually with an emphasis on their remarkable regenerative abilities, these large predatory worms are a bit nightmarish. If their regenerative capacities are like those of their smaller relations, they would seem a perfect model for a horror movie creature, and indeed they sit surprisingly high on the food chain. There are also sponges, snails, and copepod crustaceans. At a smaller scale, numerous undescribed nematode worms await study, as they do in many habitats; ostracod crustaceans, or seed shrimp, are notably diverse as well.

The abyssal denizen that is perhaps most surprising is an insect. It is quite ordinary to find insect larvae in streams and lakes, but finding them under more than a kilometer of water, which they must somehow safely traverse to metamorphose into adults and reproduce, is unexpected to say the least. Yet in the stygian recesses of Baikal, where for millions of years sunlight has been a rumor from afar, and the difference between a blistering July and the most numbing January is a fraction of a degree, one can find the larvae of chironomid midges. These are not exotic beasts like a horror movie flatworm or a golomyanka but instead prosaic little creatures that are related to the "bloodworms" commonly sold as frozen food for aquarium fish.

Most of the deepwater denizens of Baikal get their nutrition from the rain of dying, dead, and decaying organic matter that emanates from productive waters closer to the surface, just as do many deep-sea creatures. Thus, the depths are in most places something of a desert, though really there is no terrestrial habitat that provides a satisfactory analogy. Even in a desert there are cacti or other organisms that manage to perform photosynthesis and provide a foundation for a food chain. In the deep, there is just that slow rain and the occasional bonanza of a dead sturgeon or maybe a seal, likely to be quickly pounced on by specialized scavenging gammarids with keen senses for the detection of such prizes.

Except . . . just as in the deep sea, in Baikal one can sometimes find a source of energy and carbon that comes from below. Baikal has hydrothermal vents and methane seeps, and these are the basis for higher-density animal communities, much like in the ocean. So far, there do not seem to be any Baikalian equivalents of the tube worms and bivalves of deep-sea vents, which have chemoautotrophic (i.e., extracting energy from the materials flowing from the vent, without photosynthesis) bacteria living symbiotically within them, but there are free living bacteria that similarly form the basis for food chains. This is not completely unprecedented in fresh water. For example, there are vents in Lake Tanganyika that are associated with communities of microbes, but any such communities in the abyss of Tanganyika would be in waters that generally lack oxygen. Hence persistent animal communities around deepwater vents would not be possible.

The study of Baikal's vent and seep communities only got started in the last few decades as deep-diving minisubmarines became available for the direct exploration of the lake's deeps. Exploring Baikal's unknown abyss in a minisub sounds like a fantasy come to life for anyone who grew up watching Jacques Cousteau and similar programs, as many of us did. But apparently these trips require fortitude. A colleague of mine who has made descents in the Atlantic reported how tight the confines were, with two or three people crammed together and no bathroom facilities. On one trip some years back, one of the scientists developed digestive tract problems; since there is great reluctance to interrupt these expensive excursions and the ascent can easily take a few hours regardless, the atmosphere in the sub became rather thick.

The advantage of the subs is that direct, recorded observations can be made of the environment around the sub, and carefully selected samples, including live animals, can be collected

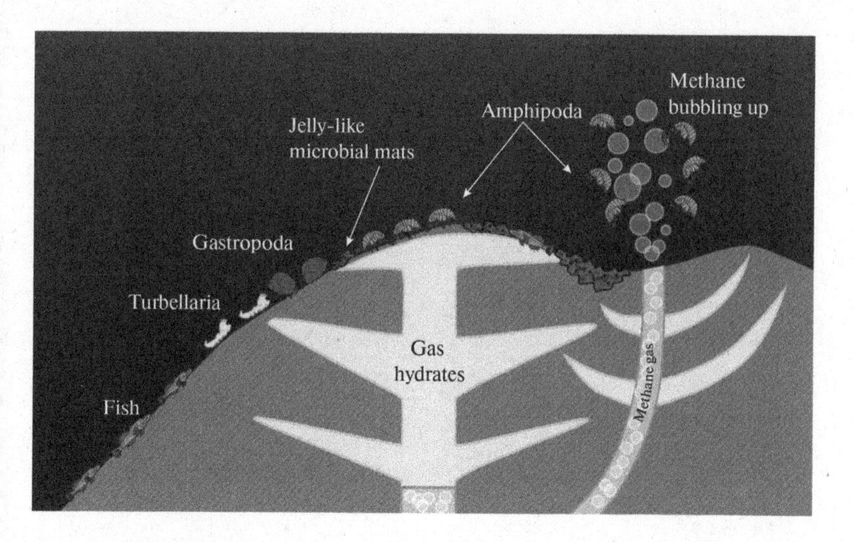

Figure 8.8
"Saint Petersburg" Baikal methane seep community. *Source:* Reprinted with minor modifications with permission from Springer Nature, from Sitnikova et al., "Trophic Relationships between Macroinvertebrates and Fish in St. Petersburg Methane Seep Community in Abyssal Zone of Lake Baikal," *Contemporary Problems of Ecology* (2017).

with slurp guns, sediment corers, and other devices. The video observations have revealed otherworldly environments quite different from the flat, deep sediment that otherwise dominates the bottom of Baikal.

At the Saint Petersburg methane seep, about 1,400 meters below the surface in the central portion of Baikal, the bottom landscape is small hills of four to six meters in height (figure 8.8). These are composed partly of ice-like, transparent blocks of gas hydrate, which are best known from the deep sea. In the ocean, they sometimes form where methane bubbles up under the terrific pressures and low temperatures of deep water. At

Figure 8.9
The abyssal sculpin *Abyssocottus korotneffi* alongside a giant planarian of the Dendrocoelidae at Baikal's Saint Petersburg methane seep. *Source:* Reprinted with minor modifications with permission from Springer Nature, from Sideleva, "Communities of the Cottoid Fish (Cottoidei) in the Areas of Hydrothermal Vents and Cold Seeps of the Abyssal Zone of Baikal Lake," *Journal of Ichthyology* (2016).

other sites, small, gurgling mud volcanoes may occur, or hillocks and tubes of bitumen, essentially an oily tar.

Mats of bacteria are common at these lake-floor oases and likely form the basis for the food chains. Many of the organisms generally characteristic of abyssal Baikal are present at these sites, though in much-elevated numbers. Thus, there are sponges, sculpins, planarian worms, annelid worms, gastropods, chironomid midge larvae, and others. Large flatworms are often notably common (figure 8.9).

While the species present overlap extensively with those of the surrounding lake-floor desert, the relative abundances of individual species, which are best known for sculpins and gammarids, may be different. They may also differ between seeps and vents. Among the sculpins at the Saint Petersburg methane seep, for example, Valentina Sideleva of the Russian Academy of Sciences reports that *Neocottus werestschagini* is the most common species. This species, however, is infrequently encountered on the lake bottom generally, or about fourteen times less often relative to other species, than at the seep. At the Frolikha Bay hydrothermal vents, located in 400–480 meters of water, a species of the same genus was overwhelmingly dominant, but this time it was *Neocottus thermalis*, a more extreme specialist in that it is only found at hydrothermal vents and is the only sculpin known to possess such an exclusive affinity. It has no obvious adaptations that restrict it to the vents so the reason for this inflexible association remains unknown. The two golomyanka species are also notably abundant at both the vent and seep sites. Up to fifteen individuals could be seen in a single video frame taken at the vent, and thirty at the seep. They were therefore several times more abundant than they are in the open lake. This pattern too is a puzzle, as they are not feeding on vent or seep organisms; instead, they are sometimes eaten by the bottom-dwelling sculpins that are abundant there.

By examining the carbon and nitrogen isotopes present in the organisms dwelling at the seep, Sideleva and a group of Russian colleagues from several institutions were able to infer whether seep organisms' nutrition originated with seep microorganisms or came from a food chain that began with photosynthetic production closer to the surface. They found that many of the bottom-dwelling sculpins, gammarids, and planarians appeared ultimately to be deriving most of their energy and carbon from the seeps. Similar results have been obtained for vent systems, although in contrast to the bottom-dwelling sculpins, golomyankas were found to derive their nutrition from the open water ecosystem, where most energy originates with photosynthesis near the surface. None of these analyses have identified any organisms that appear to be hosting symbiotic bacteria that are generating energy and carbon compounds. But all conclusions for these systems are still based on small samples. Surely there will be many more surprises coming, especially as exploration expands through the use of less expensive, remotely controlled underwater drones.

BAIKAL IN THE ANTHROPOCENE

The final chapter of this book is focused on the accelerated pace of change lately confronting ancient lakes, as our species comes to dominate this unusual little planet ever more exhaustively. Yet it would seem incomplete to depart from Baikal without saying a few words about the developments there during the last decades, particularly since they encompass major features of its massive ecosystem. Processes in Baikal are also somewhat special because although there are other ancient lakes at high latitudes, especially if one uses Stephanie Hampton and colleagues' definition, Baikal is the highest-latitude ancient lake with extensive endemism, and the only such lake covered in ice each winter.

As the largest and deepest of freshwater lakes, with a vast volume comprising 20 percent of the planet's liquid fresh water, one might expect Baikal to be resistant to change. Thus, there was a good deal of interest when comprehensive analyses began to appear in the 2000s of the sixty-year data sets collected by Kozhov, Kozhova, and Izmest'eva. These and other data show clearly that Baikal is warming and that the annual duration of ice is shrinking (figure 8.10). It is also becoming apparent that these changes are affecting the lake's organisms indirectly through effects on other physical processes in the lake as well as directly. In some cases, changes in physical processes are affecting how organisms interact with each other.

In the first major report presenting comprehensive analyses of the data collected by the Kozhov family, Hampton, of the US National Center for Ecological Analysis and Synthesis (now at the Carnegie Institution for Science), Izmest'eva, and a team of collaborators from multiple institutions reported on the biological changes that had accompanied the warming of Baikal. They found that algal mass has been increasing overall, as have the numbers of a group of widely distributed zooplankton known as cladocerans, which do well at higher temperatures. In contrast, the endemic, cold-loving *Epischurella* has been either declining slightly or stable. Owing to physiological and other differences between the different types of zooplankton, Hampton, Izmest'eva, and colleagues suggest that if these trends persist or intensify, patterns of nutrient cycling in the lake could be substantially affected, with broad ecological consequences.

In a complementary analysis of data from shallow sediment cores, an international team led by British scientists George Swann (University of Nottingham) and Anson McKay (University College London) looked at how natural and human-driven

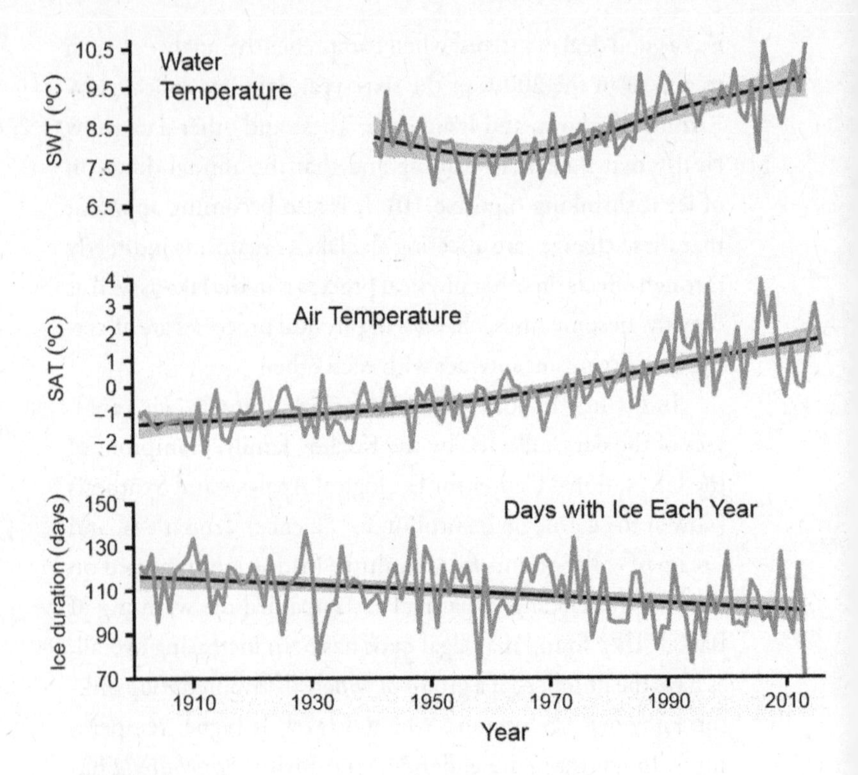

Figure 8.10

Temperature and ice cover trends at Baikal. *Source:* Reprinted with minor modifications from Swann et al., "Changing Nutrient Cycling in Lake Baikal, the World's Oldest Lake," *Proceedings of the National Academy of Sciences* (2020). Published under a Creative Commons CC BY 4.0 license (https://creativecommons.org/licenses/by/4.0).

changes have affected nutrient and chemical cycling, and ultimately changes in algae productivity. Their time frame of 2,000 years was longer, but still comparatively recent. Their most important conclusion is that since the mid-nineteenth century, the supply of key nutrients has greatly increased, from the nutrient-rich deeper waters to the nutrient-limited shallower waters where light is high and algae can be productive. They suggest that this is the result of documented increases in wind strength over the lake, which can cause more extensive "ventilation" of deep waters. The cause of increased wind strength is not yet known with confidence, but decreased ice cover along with increased air and surface-water temperatures likely contribute.

Hampton and Izmest'eva have built on these and other findings in a mathematical model of the Baikal open water ecosystem, developed with several additional collaborators including Sabine Wollrab of Michigan State University and Berlin's Leibniz Institute of Freshwater Ecology and Inland Fisheries. In the model, they seek to integrate biological interactions between organisms with changes in the physical environment. Their goal is to better understand the causes of the recent changes in seasonal patterns of algae abundance, especially in the winter. Baikal, with sunlight penetrating its clear winter ice, has traditionally had a peak in algae productivity in the winter and early spring—yet another unusual feature of this system. In the late twentieth century, these peaks were often delayed, weaker, or simply absent. The Kozhov family's data detected these patterns, which can seldom be evaluated in lakes, because of their determined sampling through the winters. The model, which takes into account *Epischurella* abundance and grazing, and considers separate populations of cold-adapted and warm-water-adapted algae, suggests that these changes in algae abundance may be largely the result of reduced annual ice cover and that if ice coverage continues to diminish the winter algae peak may disappear altogether. The model is somewhat complex, but its predicted outcomes arise at least in part from the greater ability of the *Epischurella* to suppress algae population growth by eating the algae when there is less ice cover. The model describes a "regime shift," a steplike switch from one state of a system to a different state involving a different range of variation. No model is final, and this one may evolve as our understanding of the ecological interactions evolves, but the contrast between regime shift and steady, gradual change is worrisome and even frightening. It indicates that global warming and other human-generated environmental changes may sometimes cause abrupt shifts in ecosystems that may be hard to both predict and reverse.

* * *

Lake Baikal, the largest and most ancient of freshwater ancient lakes, had its start in the time of the dinosaurs and began to take its modern form well before the appearance of our own lineage, the Homininae. Yet it only assumed its current deep and thoroughly oxygenated character in the late Pleistocene. Among its diverse endemic fauna, its gammarid amphipods and sculpins are especially well studied. Species from both radiations are uncharacteristically important in open water food chains, and also as prey for the planet's only species of freshwater seal, the nerpa. Other gammarid and sculpin species are important in

Baikal's highly distinctive abyssal vent and seep communities. As the biodiverse ancient lake at the highest latitude, Baikal is showing the direct and indirect effects of global warming on its physical and biological systems and processes. The lake may be experiencing an ecological regime shift that should give pause to creatures living in a larger yet still finite ecosystem—one that is quickly heating too.

9 PLIOCENE, PLEISTOCENE, HOLOCENE, ANTHROPOCENE . . . ANCIENT LAKES MEET MODERN *HOMO SAPIENS*

> *We want to leave a lake without pollution for our children. . . . [W]e can fight every day to care for and clean our Lake Titicaca, unique in the world.*
>
> —Anonymous, from an interview in Peru near shore of Titicaca (translated from Aymara)

When I first went to Sulawesi as a teenager in the 1980s, I spent a few days in Jakarta on my way home, visiting the extended Indonesian family I gained when my (Canadian) uncle Brent married my (Indonesian) aunt Paulina. One of my cousins toured me around Jakarta on his motorbike, and although neither of us had much facility in the other's language, we had a grand time. After I returned home, we lost touch, but I looked him up almost twenty years later when I returned to start research in Lake Matano. He too had finished his education and had begun a career in the tech field, but when the Asian financial crisis hit in the late 1990s, he had to find other work. He ended up in the timber business, buying and selling logs. In Indonesia in the 1990s and 2000s, this meant dealing with private companies, but also the military, which operated a variety of businesses. It was fascinating to hear of his experiences, for a biologist interested in conservation, but disturbing too. Indonesia's forests are dwindling, like most tropical forests, and their loss affects watersheds, lakes, and ultimately local as well as global climates. I was left with a strong sense of the economic realities and contingencies that drive the individual decisions that collectively, will have major effects on how badly the current mass extinction turns out.

As we move into the Anthropocene, the proposed geological epoch that began with the start of significant human impact on the earth's ecosystems and geology, ancient lakes are experiencing some of the same stresses that are affecting tropical forests. They are also subject to other, distinct stressors and are in important respects especially vulnerable; lakes are like islands of water in a sea of land, and their denizens are susceptible to disturbance in the same way that creatures that have evolved in isolation on islands are easily extinguished, and their ecosystems disrupted. We know that of the extinctions documented since European expansion around the globe, approximately 75 percent have occurred on islands, and islands harbor about half of the species that are currently threatened—despite covering less than 7 percent of land surface area. It has also become clear that population declines and extinction risks are generally higher for freshwater species than for other habitats. Moreover, as for some islands, the discrete boundaries typical of large lakes remove an important option available to land-dwelling continental organisms. If lake creatures are suffering from warming (or other stressors), they can rarely just move to a new home farther from the equator or at a higher altitude. Thus, global warming is expected to cause many more ancient

Figure 9.1
Lake Lanao. *Source:* Björn Stelbrink.

lake extinctions in the coming years, but other human-caused stresses have already caused extinctions. Mindanao Island's Lake Lanao provides an extreme example.

Lake Lanao (figure 9.1) is the largest lake in the Philippines. Its age is somewhat uncertain, but it is likely over 100,000 years and is widely treated as an ancient lake (see, for example, figure P.1). Lanao is not notably large, but with an area of 357 square kilometers and a maximum depth of 112 meters, it is substantial. When John Langdon Brooks wrote about Lanao in his classic 1950's synthesis, he emphasized the radiation of cyprinid fishes (minnows, carps, and kin) that completely dominate the lake's fishes, with approximately eighteen endemic species and only two species of fish from other groups. Albert

Herre of Stanford University conducted the initial studies of Lanao's fishes and found some of the cyprinids so distinct in form that he assigned them to different genera—a total of five. George Sprague Myers, also of Stanford University, was just as impressed with the diversity and uniqueness of Lanao's fishes, asserting that "the peculiar . . . lower jaw modifications . . . are approached nowhere else in the very large family Cyprinidae, which is generally distributed throughout Eurasia, Africa, and North America, and exhibit many remarkable specializations . . . [that] transcend the familial limits of all the 1,500 to 2,000 non-Lanao cyprinid species in the world."

In addition to the cyprinids highlighted by Brooks, there is a substantial array of gastropod mollusks in Lanao. Thomas von Rintelen, whom we met earlier, collected there in the late 2010s with Björn Stelbrink of Germany's Giessen University. Björn was involved in several of the studies already discussed. Alongside being an intrepid field collector, he is noteworthy for the breadth of his ancient lake expertise, having worked on diatoms, invertebrates, and vertebrates from lakes spread over four continents.

With the specimens from their fieldwork, von Rintelen, Stelbrink, Philippines scientist P. O. Naga, and two additional colleagues conducted the first DNA sequence-based analyses of Lanao invertebrates. The exact number of species in Lanao is the subject of continuing study, but the diversity is likely comparable to the fishes, and may involve one or more endemic radiations with a wide range of forms.

The tragedy of Lanao is that of the eighteen endemic cyprinid fishes, sixteen appear to be extinct, based on surveys in the late 2000s by Gladys Ismail and colleagues at Oregon State

University. (One of these colleagues was David Noakes, one of my professors when he was at the University of Guelph. He died in 2020, but left a fine legacy of former students and valuable contributions, such as the paper described here.) The situation is not entirely hopeless because it is clear from surveys conducted over the preceding decades that particular endemic species have sometimes disappeared for a time and later reappeared. This could be partly a result of fishing methods, as most of the survey work relied heavily on collections by fishers, and partly due to the low numbers of endemics. Nevertheless, it is not encouraging that dramatic increases in introduced fish species have accompanied the decline of endemic fishes; in fact, Ismail and her coworkers documented more species of nonnative fish in the lake than native! And additional introductions have since been documented. A further complicating factor is the political situation in the area, which has often involved a level of violence and insecurity that makes scientific research not only difficult but at times genuinely dangerous too. Still, further genetic work on the fishes of Lake Lanao is presently underway to facilitate identifications. One hopes that as these efforts expand, they will enable environmental DNA surveys and perhaps detect populations of some of the species currently suspected to be extinct.

The rapid decline of Lanao's cyprinids is still more regrettable because they are so little known; even preserved specimens are sparse. Herre's original collection was destroyed in 1945 during the World War II battle for Manila. A few specimens collected separately were placed in Stanford University's Natural History Museum and the US Smithsonian Museum, but they are no substitute for Herre's.

Things seem to be better for Lanao's mollusks. No extinctions have been reported, and Stelbrink and his colleagues collected a wide range of species of diverse forms in 2016 and 2017. Worrisomely, however, they observed high numbers of an invasive species, and human activities have caused deterioration of the water quality. As well, the invertebrates are so poorly known that some could have disappeared before they were ever collected and documented.

The situation in Lake Lanao is extreme, but the problems found there are widespread in freshwater ecosystems, including most of the ancient lakes explored in this volume. Before further reviewing the difficulties that our species is creating for ancient lakes and their biotas, though, it is worth thinking about why conserving these systems merits some effort. Having read this far, you may be like me and find it obvious why biodiversity is important, its value perhaps one of the "truths (*we hold*) to be self-evident," to quote a document that is famous where I live. But the worth and importance of biodiversity are not so obvious to all.

THE VALUE OF ANCIENT LAKE BIODIVERSITY

Biologists, economists, ethicists, and others from a range of backgrounds have ruminated deeply about why declining biodiversity should concern us, and why we should be willing to invest resources in arresting that decline. For myself, like many people, the issue is an emotional and moral one. I find joy and connection in the animals and plants I encounter in lakes, forests, and just my garden, and I find time in nature restorative. It also seems to me deeply immoral that we would accept the decline of the only biodiversity we know of, not just on this moist little rock, but on any planet anywhere. This view is at least partially shared by most major religions, though not discussed terribly often, unfortunately. I suspect that these rationales may prove the most important ones in the long term, but there are more practical arguments for slowing the extinction rate.

In 2020, Robert Sterner of the Large Lakes Observatory at the University of Minnesota at Duluth, along with an interdisciplinary team of Minnesota colleagues, reviewed the ecosystem services provided by large freshwater lakes, including several ancient lakes. This compendium is the first of its kind and offers a helpful starting point for considering why we should value ancient lakes, even if from a slightly selfish "What have you done for me lately?" perspective.

The complete list of services provided by earth's ecosystems, including lakes, is a long one. The European Environmental Agency's Common International Classification of Ecosystem Services (version 5.1, cices.eu), which Sterner and colleagues reference, lists ninety such "contributions that ecosystems make to human well-being." Other organizations, such as the US Environmental Protection Agency, have their own schemes. The European list was developed to facilitate a more systematic, standardized approach to mapping, valuing, and generally describing the services supplied by ecosystems. These are the major categories: provisioning of material and energy needs; regulation and maintenance of the environment for humans; and cultural—that is, the nonmaterial characteristics of ecosystems that affect the physical and mental states of people. Sterner and his colleagues decided to focus on six of the most readily quantifiable services, four from the provisioning category and two that are cultural. Their twenty-one-lake survey included the five largest freshwater ancient lakes from three continents—the African Great Lakes of Malawi, Tanganyika and Victoria; Baikal (Asia); and Titicaca (South America)—as well as sixteen large but younger lakes.

The preeminent quantified service in their analysis was the provision of fish, and this was dominated by the African Great Lakes (figure 9.2). The total catch by commercial and artisanal fishers for Malawi, Tanganyika, and Victoria was estimated at 1.3 million tonnes per year, which comprises 95 percent of all the fish commercially landed from the planet's twenty-one largest lakes, and over 10 percent of landings by freshwater fisheries overall. These fish catches are also important locally as a source of nutrition, especially in areas of high food insecurity. At one time, fisheries were estimated to provide 70 percent of the total animal protein in Malawi, which is landlocked, although that percentage has decreased. Lake Victoria is distinct because so much of its fish harvest is destined for export, mainly of the introduced Nile perch (which I have seen for sale at my local grocer), and because of the exceptional scale of its fisheries. These estimates do not include subsistence fisheries, which may be substantial and provide valuable nutritional benefits locally. Another major service of lakes is supplying water for drinking, irrigation, and other uses. Biodiversity is relevant to water itself because living organisms and ecosystems frequently play a major role in water quality. Wetlands, for example, can remove impurities and toxins, whereas some algae and other microorganisms may introduce them, especially with the addition of agricultural fertilizers. The ancient lake providing the most water for agriculture was Titicaca, with no other lakes at

Figure 9.2
Fish for sale in a market near Lake Malawi.

all close. Sterner and colleagues were unable to quantify water removed for drinking that did not go through public treatment (treated water was included), but this is likely to be a vitally important service for many.

Lakes are popular locales for tourism and recreation almost wherever they are found, and in lakeside communities, residents often care passionately about "their" lake. Many of us have seen this ardor for a local water body in members of our own communities and families. The analysis by Sterner and colleagues used social media data to evaluate tourism and recreational activity. They documented substantial activity at all the ancient lakes relative to the rest of their data set, although the social media data offer only a crude first approximation of the scale and value of this service. The one ancient lake for which reasonably reliable quantitative data were available, concerning the financial value of tourism, was Baikal. Baikal has become a major tourist attraction. The economic value of Baikal's tourism was estimated at US$266 million per year.

Although not a principal focus, Sterner and colleagues were able to collect some data relevant to how ecosystems support scientific investigation by counting the number of scientific papers associated with each lake from 1900 to 2018. Baikal had the highest value, followed by the African Great Lakes, with Titicaca well behind. My impression is that scientific research on the lakes is rapidly growing and broadening so these estimates too are probably conservative. And studies published so far are likely to be the tip of the iceberg.

The Minnesota team was able to summarize and analyze data on other services supplied by lakes, such as functioning as a transportation corridor, or providing water for electric generation or cooling a power plant. Although these activities could affect biodiversity, likely negatively, they rely on living things very little. In contrast, numerous other services are potentially important and more directly linked to biodiversity, but difficult to quantify at present. These include services related to "aquaculture, mediation of wastes, regulation of flow, biodiversity protection, climate regulation, nutrient cycling, and spiritual and heritage benefits." Aquaculture, in particular, is currently increasing at a tremendous rate on several of the lakes.

One underutilized and undervalued service bears particular emphasis: the provision of new drugs along with molecules possessing scientific, medical, and even industrial utility and value. It is now well-known and appreciated that many drugs originated in natural systems. Some famous examples include aspirin, which was originally derived from willow bark, and digitalis, a drug that originated with the foxglove plant and is used to treat heart disease. Penicillin and morphine are others. Perhaps just as important, key molecules for biotechnology and scientific research continue to emerge from microbiology. Most notably, the CRISPR Cas9 and related systems, which are revolutionizing diverse areas of biology by enabling much more precise genome editing, are derived from the immune systems of microbes. These molecules and systems evolved to protect bacteria and archaea (a lineage of structurally simple microbes similar to yet distinct from bacteria) from viruses and other parasites. Thus, they are a product of coevolution between the bacteria and their parasites. A distinct but equally famous illustration is Taq DNA polymerase, the crucial enzyme in the polymerase chain reaction (PCR), the DNA amplification process now used throughout biology and medicine—including in an important diagnostic procedure for COVID-19. Taq polymerase was originally derived from *Thermus aquaticus* bacteria living in the heated springs and lakes of the Yellowstone National Park in the United States. The enzyme's ability to function at high temperatures, which is essential to its role in

PCR, is an evolutionary adaptation to the torrid conditions in which the bacteria live.

Research continues to generate new discoveries and applications in drug discovery, molecular biology, and related fields that rely on the adaptations present in the diversity of nature. Given that a large proportion of extant species have not even been formally named, and many more have received little more than a name, there is clearly an immense reservoir of biochemical tools awaiting discovery. It is both sobering and awe-inspiring to consider the abundance of molecular systems that have evolved over billions of years as organisms have adapted to the changing physical world and each other, finding ways to overcome extremes of temperature, acidity, pressure, and toxins physical and organic. All the while they have maintained essential physiological processes and the machinery of reproduction, and solved endless additional challenges. Hence the information contained in the living world is unimaginably vast.

The literature on bioprospecting for drugs and other valuable molecules has to date emphasized biodiverse terrestrial systems such as tropical forests, and in the marine setting, coral reef ecosystems. But work is beginning to appear from ancient lake systems. As aquatic islands of unique evolution, ancient lakes must possess solutions to problems that face us today or that may confront us 100 or even 500 years from now. And with the revolutions taking place in genomics and related fields, including our ever-increasing computational abilities, humanity's ability to find and recognize these valuable products of natural selection grows more powerful almost by the day. Surely this immense natural capital should be more highly valued, and we should be more conscious of it.

In considering cultural services, we have discussed tourism because it can be quantified, but of course the cultural significance of iconic places, landscapes, and organisms as well as the intertwined traditions have deeper value. I live in North America, where we lack lakes quite like Tanganyika and Baikal. Nevertheless, we have analogies, and it may be helpful for readers from my original neck of the woods to imagine the cultural impact if, say, the last patch of British Columbia's old-growth forest were to burn up, or if the orcas of the North Pacific were to succumb to their many stresses and disappear entirely. Unfortunately, addressing the cultural importance of ancient lakes adequately is beyond what I can manage here. But some of the quotations at the beginning of each chapter can perhaps convey a small sense of the significance of the lakes to the cultures linked to them.

EXTINCTION, ECOLOGY, AND HIPPO

Threats to biodiversity are sometimes summarized with the acronym HIPPO, which is short for *h*abitat destruction, *i*nvasive species, *p*ollution, *p*opulation growth (human), and *o*verharvesting. The order is often described as roughly reflecting the order of importance of these factors, though that will vary among localities. The acronym is a handy one, and particularly helpful for terrestrial habitats, but like any simplification of a multifaceted issue it has shortcomings, and more so when applied to fresh water. The biggest problems are the overlap and interdependence of the categories, especially population, which affects all the others, and where climate change fits. In an ambitious review paper with a large, international team, Stephanie

Hampton and collaborators, whose work we looked at earlier with regard to Baikal, used a related but distinct scheme to summarize the ecological changes taking place in ancient lakes. I roughly follow their approach in this section.

The elephant in the room, or perhaps the HIPPO in the lake, for any twenty-first-century discussion of environmental issues is bound to be climate change, aka global warming, aka global heating; every term has its enthusiasts and merit. Certainly the lakes are warming. Hampton and colleagues compiled summer temperature data for fifteen of the twenty-nine lakes in their survey and confirmed warming for all of them, including seven of those emphasized in this volume. Almost all values fell between about 0.1 and 0.8°C *per decade*. This rate of increase, every ten years yielding differences large enough that a person could feel them on their skin, is somewhere between jaw-dropping and horrifying, and maybe both. *Ten years!* As we have seen, the lakes have experienced changes before, sometimes fast and in other instances more gradual, but the changes have had consequences, and the current rate of change is very, very rapid.

Temperature changes of this scale can affect organisms directly, especially as the decades scroll by and warming accumulates. Some of Baikal's species that are adapted to a narrow and frigid range of temperatures come to mind. At the other extreme, organisms in tropical lakes may be living at their physiological maxima and also suffer as temperatures elevate. But as for terrestrial environments, changes in the temperature regimes of lakes may have their greatest effects on ecosystems indirectly through other changes to the physical environment. The most obvious one is how long ice cover lasts each year as well as ice thickness and extent; these effects are seen on Baikal at present, as noted in the previous chapter. They affect the Caspian Sea too.

A less visible but ecologically highly consequential effect of warming is on how lakes are layered by temperature, or to use the more formal term, *stratified.* In warm lakes, warmer water is layered on top of colder water, and with a stable environment, the deep water loses oxygen and becomes inhospitable to any oxygen-dependent organism, including almost all animals. The upper layers have their oxygen replenished by algae as they perform photosynthesis, whereas in deep water there is not enough light for this process and only oxygen-depleting activities occur, such as the breakdown of organic materials floating down from above. With little mixing, the oxygen gradient can become extreme. Such stratification is common in tropical lakes like Tanganyika, and with warming, thermal stratification can increase further.

Andy Cohen, who made appearances in earlier chapters, has worked with a team of collaborators from several institutions to analyze data from a series of shallow Tanganyika sediment cores, and evaluate the links between temperature, stratification, and biological processes over the last 1,500 years. Warmth-enhanced stratification in Tanganyika has been suggested to correlate with reduced productivity of algae because the strengthened stratification reduces the amount of mixing between the shallow waters where algae live and photosynthesize, and the deeper waters that contain key nutrients needed by the algae. This can result in reduced productivity from the nutrient-starved algae, and since they are the basis for the rest of the food web, reductions in abundance of the animals that ultimately depend on the tiny photosynthesizers.

The patterns in the sediments studied by Cohen confirmed temperature increases in lake water that were accompanied by reduced algal productivity, particularly in the twentieth century. Corresponding declines were documented in the fossils of bottom-dwelling snails in the samples from intermediate depths, where they should be most affected by a shallower layer of well-oxygenated water (figure 9.3). Calculations by Cohen and his collaborators show that the area of Tanganyika's bottom that has enough oxygen to support most animal life is shrinking rapidly as oxygenated waters penetrate less deeply each decade. The reduction in habitable lake floor since just 1946 is estimated at 38 percent.

Fish from Tanganyika comprise about 60 percent of the animal protein consumed in the lake's vicinity—a really immense proportion if one compares it to typical North American diets. Thus, shrinking catches are a cause for deep concern. Cohen and colleagues' analyses show, somewhat unexpectedly, that the declines in catches are much less a result of overfishing than had been thought. Instead, the catch declines are associated with processes linked to climate change, most likely owing to reduced productivity. In fact, the declines started before large-scale fishing got underway on the lake, although this falling off has surely been exacerbated by overharvesting. One way to put the scale of the sediment data in perspective is to think of them in terms of human history. The data in figure 9.3 mainly start shortly after the fall of the Roman Empire, just as the Sui dynasty was becoming established in China and the Arab conquests beyond the Arabian Peninsula were about to begin. The data set continues, essentially uninterrupted, to the start of the twenty-first century.

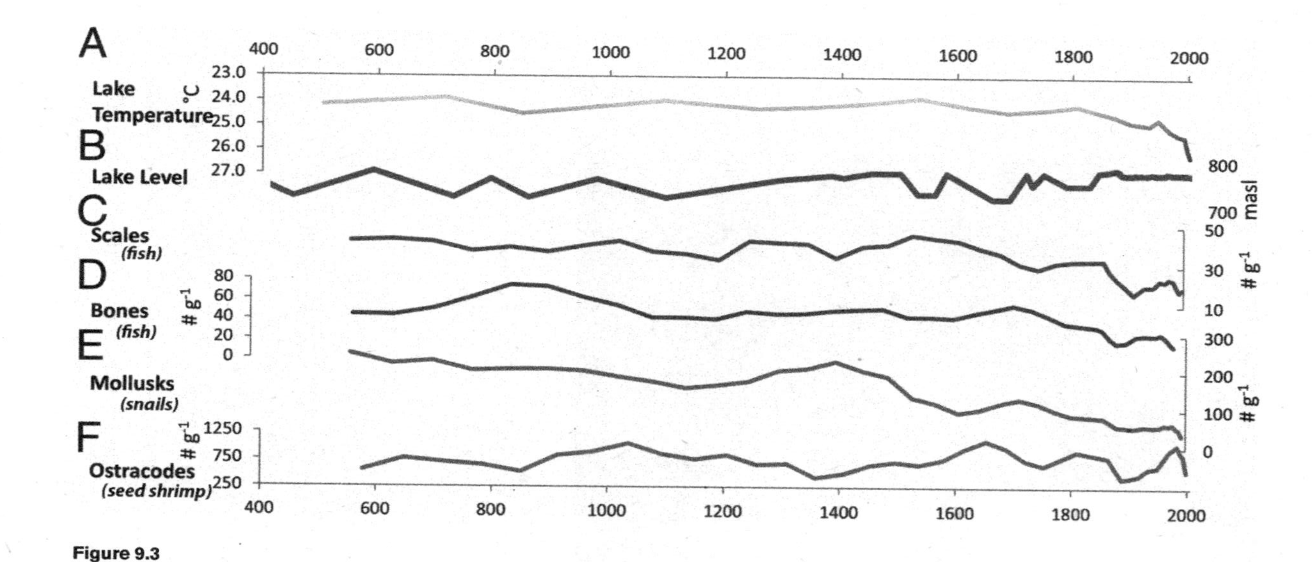

Figure 9.3

Changes in lake temperature, level, and faunal abundance (scales and bones for fish) from about 1,500 years of Lake Tanganyika sediments. Note that the temperature increases toward the *bottom* of the page in "A," and the numbers decline for nearly every organism in the twentieth century, as temperature change accelerates. *Source:* Reprinted with minor modifications with permission from the National Academy of Sciences of the USA, © 2016 National Academy of Sciences, from Cohen et al., "Climate Warming Reduces Fish Production and Benthic Habitat in Lake Tanganyika, One of the Most Biodiverse Freshwater Ecosystems," *Proceedings of the National Academy of Sciences* (2016).

Figure 9.4
Tumaini "Tuma" Kamulali sub-sampling Lake Tanganyika cores at LacCore facility, University of Minnesota. Source: Tumaini Kamulali.

The analyses by Cohen and his colleagues focused on samples from Tanganyika's northern basin, which is generally better studied than the lake's southern reaches. But Tanganyika is our planet's longest lake, at about 670 kilometers from its northernmost to southernmost points, and it would not be surprising if ecological processes varied over such a scale. This possibility was explored using sediment cores from Tanganyika's southern basin by Tumaini "Tuma" Kamulali (figure 9.4), a PhD student in Cohen's group who also holds an appointment with the Tanzania Fisheries Research Institute. I met Tuma at the 2022 Species in Ancient Lakes conference in Kigoma, Tanzania, where he won an award for best student presentation. His account of the sacrifices involved in conducting his research and pursuing his studies was affecting; having traveled to Arizona to begin his analyses before the COVID pandemic started, he was unable to return to his wife and new baby for a total of four years owing to pandemic travel restrictions and complications. Isolation and sacrifice are regrettably commonplace among foreign graduate students, but even so his experience was extreme.

Overall, the results for the southern basin also indicate that changes in climate have mediated the lake's productivity, which has declined, but the southern basin is often less stratified than the northern, and the links between temperature and productivity are less consistent. Kamulali and colleagues also found unexpected evidence that although deeper reaches of the southern basin frequently lacked oxygen, there were several episodes in the sediment records during which bottom-dwelling species were present at great depths. Hence there appear to have been periods of elevated oxygen even along the lake bottom, possibly when oxygen-containing river waters that were cooler or more turbid (and denser) flowed into the lake and down its steep sides. I find this idea intuitively appealing because I encountered a similar phenomenon in Lake Matano while swimming along the outlet of a cool stream as it entered the lake. I could feel the water plunging, which was disconcerting as I looked down into the darkness and felt a little tug toward the depths.

Climate change is expected to affect lake levels as well, which as we have seen have often changed by hundreds of meters, with dramatic effects on physical and biological processes. The effects may be most severe in lakes located in warm, arid areas where a shift in the relative rates of evaporation and precipitation could cause rapid changes in that balance.

The effects of climate change may be compounded or even superseded by dams on inflows, outflows, or both. These are present in most ancient lakes. In the Malili Lakes, for example, when I visited as a teenager there was already a dam on the outflow, the Larona, which was built to provide electricity for the nickel mine where my uncles worked. Another dam was added in the 2000s on the Petea River, which ultimately influences the water entering the Larona through the intervening Lakes Mahalona and Towuti. Depending on policies at any given time, these structures may result in higher or lower lake levels, but the effects of dams on lakes are often further complicated by the withdrawal of water from rivers feeding into a lake. The most extreme instance of such effects is the five-million-year-old Aral Sea.

Located in Central Asia and stretching across the border between the nations of Kazakhstan and Uzbekistan, the Aral Sea was once the fourth-largest lake in the world. There have been human settlements on the Aral's shores for thousands of years, and its levels have fluctuated substantially over the millennia. With no outflow rivers, it has also been slowly filling in for approximately 140,000 years. The Aral's situation changed beginning in the mid-twentieth century when humanity's impact entered a catastrophic phase.

In the 1960s, irrigation to support cotton was stepped up and massive amounts of water were diverted before they ever entered the sea, rapidly changing the net water budget

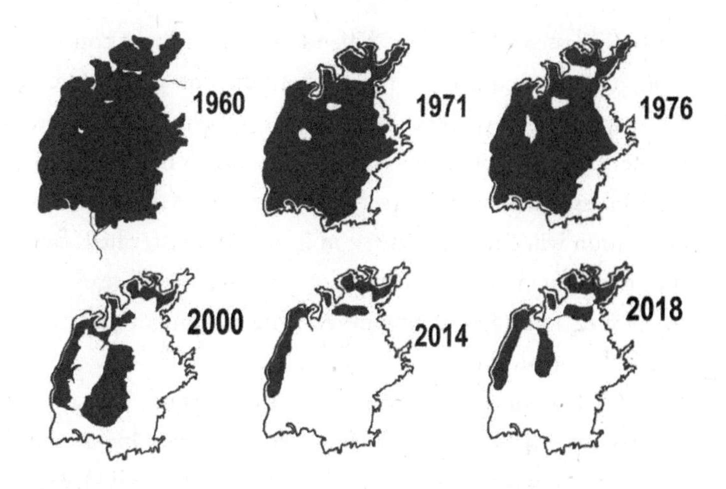

Figure 9.5

The shrinking of the Aral Sea. The dark coloration indicates water. Note that the two internal water bodies shown in 2018 were shallow and temporary. *Source:* Reprinted with minor modifications from Aladin et al., "The Zoocenosis of the Aral Sea: Six Decades of Fast-Paced Change," *Environmental Science and Pollution Research* (2019). Published under a Creative Commons CC BY 4.0 license (http://creativecommons.org/licenses/by/4.0).

into a severe deficit. With the 90 percent reduction in water volume that followed, the sea was not just reduced in depth; as it got shallower it shrank, fragmenting into several smaller water bodies separated by marsh and desert (figure 9.5). Hundreds of thousands of people were displaced from their homes, and chronic disease and even death rates rose with exposure to agrochemicals and toxic metals from the uncovered sediments. The region's terrestrial ecosystems also suffered, but the remaining waters of the sea suffered more, with salt concentrations coming to exceed those of typical seawater, whereas previously the sea had been brackish. While researching this topic, I was reminded of a remark from a cotton farmer I used to know, as cotton is grown where I live. When I asked him about cotton's reputation for requiring a lot of chemicals, he said, "If you aren't spraying it, you're thinking about spraying it." Cotton is a demanding crop, generating export dollars in places that can well use them, and it is perhaps unrealistic to think that its cultivation will end any time soon in the Aral watershed. But the environmental price is terrible; accounts of Aral read like the plot of a postapocalyptic movie. The suffering, though, is real and heartbreaking.

The news out of the Aral Sea is seldom good, but some reports have been more upbeat in recent years. One of the remaining fragments of the sea, toward the north and known as Small Aral, has experienced rising water levels with the building of dams to retain its water. Its salinities have become more moderate, and its fishing economy has begun to return along with some native fish species. The Aral Sea will probably not be what it was prior to 1960 anytime soon, and improvements have not been consistent, but further progress is possible.

After climate change, eutrophication may be the most direct and severe ecological harm visited by human actions on ancient lakes. Eutrophication mainly occurs when agricultural fertilizers enter lakes, but it may have other causes including untreated sewage and detergents that contain phosphorus. These substances, which in the case of fertilizers are designed to cause plant growth, enable the rapid expansion of algal populations, especially blue-green algae (also known as cyanobacteria). The algae often go through die-offs, causing oxygen depletion as they decompose, and particularly in the case of blue-green algae, may also produce toxins. Such processes are frequently responsible for the fish kills seen at local levels in many streams, rivers, and lakes. If eutrophication becomes widespread in a lake, it can even cause extinctions, either on its own or as one of a suite of stressors. In addition, cloudy water can make choosing a mate of one's own species more difficult for visually oriented organisms, as discussed in chapter 4—potentially leading to the disappearance of species.

Local eutrophication has been documented for most ancient lakes around human settlements and other sources of nutrients. For example, alarming reports emerged from Baikal in the 2010s concerning the effects of coastal eutrophication on the magnificent underwater sponge forests as well as other bottom-dwelling animals and plants found in shallower waters. This contrasts with the much healthier condition of offshore waters and organisms, although changes are taking place offshore too, as we have seen. The deterioration of water quality, some of it related to eutrophication, may have played a role in the extinctions in Lake Lanao too. Eutrophication is particularly extensive and intense in some parts of Lake Victoria, which is one of the lakes for which atmospheric deposition, in which nutrients arrive in the wind, is a surprisingly large source of the nitrogen and phosphorus that cause algal blooms. An estimated 55 percent of Victoria's phosphorus and 14 percent of its nitrogen arrive from the atmosphere. These atmospheric effects are most important in lakes that have large surface areas and limited catchments (the area drained by the rivers and streams that feed into a lake). The nitrogen entering lakes from the atmosphere derives from fossil fuel–burning power plants, vehicle emissions, and agriculture. Nitrogen deposition is projected to increase greatly in coming years—although one hopes that these projections will change as fossil fuel use peaks and declines.

Eutrophication, like virtually all human-caused stresses of ancient lakes, is made worse by human population growth and the associated changes in land usage. At least five of the ancient lakes that have been the focus of this volume host major urban centers with more than 300,000 people. The Caspian Sea has the most, but all the African Great Lakes have major centers along them ranging from Victoria and Tanganyika with four substantial cities each to Malawi with one. Baikal has two, and the tragic Aral Sea has four, while Biwa has high population densities and major population centers just downstream. Baikal, however, has much higher numbers of people along its shores than indicated by its permanent population owing to a large increase in tourism in the twenty-first century, much of it from China. The situation at Ohrid, like Baikal also a UNESCO World Heritage Site, is similar.

In the coming decades, population pressures from permanent residents may diminish for Baikal, Biwa, and Ohrid if national patterns are reflected locally. This will certainly not be true for the African Great Lakes, as the nations of East and Central Africa are expected to have some of the highest rates of population growth on the planet. Additional urbanization can be expected to accompany population increases as well as agricultural expansion and intensification. Population growth and land use changes will likely also contribute to increases in a wide range of other pollutants, ranging from metals to plastics to organochlorine and organophosphate compounds. Of course, these and other pollutants are not confined to areas of high population density, and some are often elevated around mines in particular. Petroleum extraction is planned or has begun for some of the African Great Lakes, with potential for severe pollution.

SPECIES INTRODUCTIONS, HARVESTING, AND INTERACTIONS WITH OTHER IMPACTS

The introductions of nonnative species constitute one of the most serious problems that human activities have caused in ancient lakes, potentially harming native species in a variety of ways. The main possibilities are that aliens may directly consume native species, outcompete them for precious resources, alter their habitats in harmful ways, or infect them with parasites for which the natives have not evolved defenses. Unfortunately, nonnative species have become established in every lake discussed in this volume. Lake Matano, the ancient lake where I have mainly worked, has had established introduced fish for decades, some over a century, and continues to experience new arrivals. When I first traveled there to do research in 2000, my collaborators and I observed several species that were already ensconced, including carp, tilapia, *Clarias* walking catfish, and snakeheads. All are commonly introduced across Southeast Asia and elsewhere, often through aquaculture escapes or deliberate stocking as food fish. We also observed an African cichlid, a single individual, that seemed likely to have come from the aquarium trade, and a tambaqui, a mainly herbivorous member of the same South American family as the piranha. The tambaqui, or blackfin pacu, is popular for aquaculture and sometimes kept in aquaria.

Although we were in Sulawesi to do basic research, we became concerned about these new arrivals and felt we should try to prevent further introductions. Our efforts included a short, informal guide to Matano's fishes, in English and Indonesian, community presentations, and later a poster campaign supported by the mining company and others to inform people about the danger of introduced species (figure 9.6). My friends and colleagues Fadly Tantu and Jusri Nilawati of Tadulako University made many additional efforts to communicate the importance of native Matano and Towuti species along with the threats to them; they deserve much credit for this work. But preventing introductions is a difficult enterprise, and our efforts were not enough, sadly.

In 2010, Fabian Herder and others documented yet another Matano invader: an aquarium species known as the flowerhorn cichlid or luohan. This creature is a human-generated hybrid of new-world cichlids and is now established in the wild in several Asian countries. In Lake Matano, it spread at an impressive and disturbing rate, and could be observed all the way around the lake by mid-2012. Subsequent reports suggest it continues to do well and feeds widely on native species including crabs, shrimp, snails, and fish, and overlaps in diet with several native fishes. Some research has begun into possible control methods. The possible effects of the flowerhorn include declines in a native goby as well as five snail species, the latter newly described in 2023 from collections made from 2003 to 2005, that appear already to be extinct. At the same time, the story is an instructive one. This is a lake next to a massive surface mine, where two dams have been added to the system in addition to pollutants and sewage, and in a region where climate change is making critical monsoon rains less predictable. Yet an indirect effect of mining and increased human density may pose the greatest threat to endemic biodiversity: the careless release of our pets.

Since Herder and colleagues first reported them in Matano, flowerhorn cichlids have also become abundant in nearby

PERHATIAN !

Awas! Melepaskan jenis ikan baru dapat mengancam kelestarian spesies ikan asli Matano!

Danger! Introduced Fish Threaten Lake Matano Species

Dua jenis ikan yang hanya ditemukan di Danau Matano : Opudi Ekor Bundar (gbr.kiri), Opudi Ekor Lancip (gbr.kanan).

Two of the fish species found only in Lake Matano: Roundfin Silverside (left), Sharpfin Silverside (right).

Danau Matano memiliki banyak spesies ikan yang unik. Ikan asing bisa menjadi pemangsa, pesaing makanan dan mendatangkan penyakit pada ikan lokal.

Lake Matano has many unique species. Foreign fishes could eat them, eat their food or give them desease.

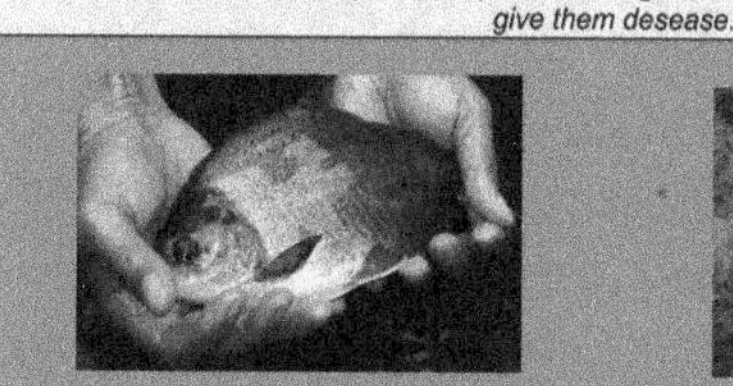

Dua spesies ikan asing ditemukan di Danau Matano: spesies ikan keluarga Piranha dari Amerika Selatan (gbr.kiri) dan Ikan Cichlid Afrika (gbr.kanan).

Two foreign fish species caught in Lake Matano: a South American member of the Piranha family (left) and an African Cichlid fish (right).

Bagaimana bisa melindungi?

- Jangan melepas ikan aquarium ke dalam danau atau membuang mereka di selokan. Jika tidak lagi menginginkan mereka, harus dibinasakan.
- Usahakan untuk tidak membiarkan ikan aquarium lepas dari tempatnya.

Terima kasih!

How can we help?

- Never release aquarium fish into the lake or put them into the toilet. If you don't want them, they must be humanely killed.
- Try not to let aquaculture fish escape from cages.

Thank you!

Sponsored by/ Disponsori oleh :
Universitas Wisconsin-Simon Fraser-Tadulako Fish Research Team; Soroako Diving Club; Limnologi LIPI; PT. INCO

Figure 9.6

The poster we made (with Fadly Tantu, Suzanne Gray, and Peter Hehanussa) in the 2000s to discourage introductions in Lake Matano.

Lake Mahalona and have appeared in the more distant Lake Poso, which is in a different drainage. In addition, a Lake Malawi species, the "golden cichlid," *Melanochromis auratus*, has become more abundant in Poso than flowerhorns, which are not so ubiquitous as in Matano; overall, the number of nonnative species in Poso is greater than the number of native fishes. It is also noteworthy that Poso may have already experienced extinctions of natives, as one goby and two ricefish have not been recorded since the 1980s.

The most famous ancient lake invasion was not at all accidental: the Nile perch of Lake Victoria. The Nile perch is a large, carnivorous fish, which can reach a massive 200 kilograms and almost 2 meters in length. It is much valued for the quality of its flesh and is native to several African rivers and major water bodies, but not to Lake Victoria. It was introduced to Victoria in the 1950s and 1960s by fisheries officials seeking to produce better and more valuable food fish from the lake. Reading accounts of this process, I learned that Geoffrey Fryer, a prominent cichlid biologist, had vigorously opposed this introduction, though to no avail; it reminded me of Peter Hehanussa and I trying to talk INCO out of building a dam at Lake Matano and later an expanded road network, with little effect too. The Nile perch has sometimes been credited with the extinction of tens or hundreds of species of Lake Victoria's haplochromine cichlids in the late twentieth century. In a synthetic account of events in Lake Victoria that appeared in 2021, however, Vianny Natugonza of Uganda's Busitema University, together with a team of colleagues from four African and European nations, suggest there is a good deal more to the story.

After carefully reviewing the sequence of events in the lake and two associated smaller lakes, Natugonza and colleagues find that the initial major wave of declines in native cichlids took place in the first half of the twentieth century and into the 1950s as a result of overfishing, which gradually got worse during this period. The decline was mainly of two native species of *Oreochromis* cichlids, and less so the megadiverse haplochromines. Subsequently, two nonnative species of *Oreochromis* (aka tilapia) were introduced along with the Nile perch in the 1950s. The introduced cichlid species tend to better tolerate heavy fishing and possibly Nile perch as well, so there were potential fisheries advantages to them.

Lake levels increased in the early 1960s with heavy rains, causing inshore habitat changes that may have stressed haplochromines. Catches of the introduced *Oreochromis* went up in the late 1960s and 1970s, with Nile perch catches following later in the mid-1980s. A haplochromine-focused trawl fishery was also established in the 1970s, and at the same time eutrophication increased rapidly owing to changes in human activities around the lake. Water transparency declined with eutrophication and so did dissolved oxygen, as the algal community became increasingly dominated by blue-green algae. In the early 1980s, the haplochromine cichlids went through a catastrophic collapse: 40 percent of the estimated five-hundred-plus species of haplochromines disappeared. Fish-eating haplochromines suffered disproportionately high declines and extinction rates. They are likely to be competitively inferior to Nile perch at preying on fish owing to their modified pharyngeal jaws—a second set of jaws present in the throat that help with prey requiring intensive processing such as plants and hard-shelled mollusks. Cichlid pharyngeal jaws may have facilitated the group's massive radiations—but they are a hindrance for preying on fish. Nile perch and *Oreochromis* landings continued to increase.

In 1989, the highly invasive South American water hyacinth plant became established. As it spread around much of the lake's shoreline, it often caused low oxygen levels beneath it (and favored air-breathing lungfish!), but the weed may also have provided a refuge from the Nile perch for some haplochromines. Climate change likely contributed to low oxygen levels in deeper waters, as in Tanganyika. In the 1990s, the Nile perch stock declined under heavy fishing pressure, and late in the decade, the water hyacinth was brought under better control. The least anticipated development of the 1990s, however, was that haplochromine numbers experienced a substantial resurgence—one that has continued. The complete haplochromine diversity has not come back: there are fewer species, they are less ecologically diverse, and total abundance is not the same. In addition, several of the returning haplochromines that have been carefully studied exhibit ecological and morphological changes, which could reflect evolutionary adaptation as well as developmental or physiological responses. Possibly hybridization has been involved, as it was during the haplochromine radiations. Regardless, the overall status of the haplochromines is much improved.

What is clear from this account is that the haplochromines experienced multiple severe stresses in quick succession and even concurrently. And while Nile perch certainly ate cichlids, changes in cichlid numbers only partially mirrored Nile perch abundance. Thus, it is unlikely that haplochromine extinctions and declines were entirely an effect of Nile perch, and there may have been interactions among factors, such as eutrophication making haplochromines less able to evade their invasive predator. The direct effects of eutrophication on hybridization are also well-known, as discussed earlier. On a more positive note, it is apparent from the recent developments in Victoria that an initial decline in native species is not necessarily permanent if the stresses can be removed. Extinctions are forever (barring extraordinary interventions), and an ancient lake with fewer stresses is unlikely to be the same as it was before modern humans, but some species and ecosystem services can return. This has been the experience in other, younger lakes too, for example, in central Europe, where extinctions took place, notably of whitefish, but eutrophication has been addressed and some species and ecosystem services have come back.

PROSPECTS

So what is coming next? And what can we do to slow the rate of extinction and ecosystem degradation in ancient lakes?

As we have seen, ancient lakes are much like islands, with high numbers of endemic species and a worrisome vulnerability to human-caused stressors. It is clear that more global warming will occur in even the best-case scenarios and more invasive species will arrive, or invasives already present will increase in numbers and impact as they adapt to local conditions or conditions change to suit them. There is therefore no likely future in which some major stresses do not increase for every ancient lake, even lakes like Biwa and Ohrid that are located in relatively wealthy and politically stable regions. In addition, lakes in areas where populations are growing and projected to grow further tend to be in locations where resources for conservation are constrained, especially the African Great Lakes and Titicaca. It is hard not to be concerned about the medium-term prospects for Victoria.

There are reasons for hope as well. For some of the worst human-caused stressors, it is obvious what needs to be done, the needed steps have direct benefits to people, and consensus in support of the necessary actions is growing. While only modest steps have been taken so far on climate change, many societies are much closer to a consensus in support of effective action than was the case at the end of the twentieth century—when *scientific* consensus already existed—and the needed technologies have become vastly cheaper. Also encouraging, eutrophication has been reversed in some lakes in Europe and North America and the necessary policies are well-known; some of these will reduce other pollutants too.

Concerning introduced species, a reduction in the rate of introductions is feasible with better regulation of the pet and aquarium trade, aquaculture, and other risky activities, though we are further from a societal consensus on the control of introductions, even in North America. The expense and difficulty of controlling or eliminating invasives once established is also cause for deep concern (I am especially worried for Sulawesi's lakes) and a strong argument for prevention. Still, there have been notable successes in the removal of invasives from terrestrial islands, sometimes leading to the rapid growth of endemic populations. Thus, successful models exist, and new approaches continue to be developed. Even so, the management of current and future invasives will be a continuing task, and a difficult one.

Fisheries management is never easy, but there are enough success stories to know that it is possible. For example, protected areas in which fishing is prohibited have proven tractable and effective in some marine systems, and their use is growing. Many ancient lakes are large enough for such spatial approaches to be viable, and they are being implemented in some, such as Tanganyika.

For lakes that extend across international borders such as Titicaca and the African Great Lakes, coordination between nations is important to effective management, but this is rarely easy or straightforward. Nongovernmental conservation organizations can play key roles in such efforts, and they have begun to do so in some cases, with Lake Tanganyika again providing an example.

A key step will be a more explicit recognition by international bodies and major conservation organizations of the value of ancient lakes as threatened hot spots of biodiversity—together with prioritizing their conservation. There is growing appreciation of the importance of freshwater biodiversity, the many threats to it, and the value of large lakes, a category that includes several ancient lakes. National and international reports and publications on these matters have been appearing regularly in the 2020s.

Beyond recognition and prioritization, money and expertise will be required, as for most worthwhile endeavors. Money will need to come from those who have more of it, hence wealthier countries, but it will need to be spent in partnership with people living around the lakes such that they benefit and traditional knowledge can be engaged. It is also crucial to invest in augmenting scientific and management expertise in the nations surrounding the lakes, and to support those scientists and managers. The rising frequency of publications in aquatic ecology, management, and evolution by researchers from some of the less wealthy ancient lake nations is a trend that deserves further encouragement, investment, and support.

I think most of the positive steps outlined here (and others as well) will be taken. Several decades from now—five? nine? more?—most of the lakes may be starting to emerge from the difficult period they are currently in—one that is likely to get worse in the near term. What is harder to know is how soon we will take those needed actions, and how much time we have. Will we do too little too late? Will most lakes emerge from this period of trauma having suffered numerous extinctions, with their ecosystems permanently and severely degraded, and temperatures still rising? Or will we act sooner and effectively enough that their endemic faunas and floras largely persist, ready to bounce back, with the major features of their ecosystems also poised to return and warming slowed? I am not sure I am optimistic, but I am hopeful—hopeful enough to have written this book. There is much to do.

* * *

Ancient lakes are like aquatic islands in a sea of land, and like islands, can easily suffer extinctions with human disturbance. Emotional and moral justifications for maintaining lake biodiversity are bolstered by documented ecosystem services, including the provision of fish protein in protein-deficient areas, clean water, and tourism. The value of biochemical adaptations for drugs and industry is likely underestimated. Human-generated changes that threaten ancient lakes include global warming (in tropical areas too), water diversion, eutrophication, species introductions, and overharvesting. These can interact. The stresses on most lakes will likely increase in the near term, but some may later diminish. With foresight and timely action, it is possible that much ancient lake biodiversity can endure the current difficult moment.

Acknowledgments

I have had a great deal of help in putting this book together and the work that necessarily came before. Brent, Carol, Jack, and Paulina Thompson, along with Perry and Sharon McKinnon, helped me get started on ancient lakes. Andrew McKinnon, Mike Pauers, Irene Raphael, George Savage, and Ian Timberlake read the whole manuscript and provided helpful critiques as well as encouragement. Rebecca Asch, April Blakeslee, Valentina Burskaia, Rod Campbell, Bob Christian, Heather Dail, Suzanne Gray, Bob Green, Carolyn Green, Stephanie Hampton, Todd Hatfield, Steve Heard, Claudia Jolls, Tuma Kamulali, Trip Lamb, David Marques, Tara McKinnon, Joana Meier, Nare Ngoepe, Jusri Nilawati, Oliver Selz, Maria Servedio, Alena Shirokaya, Björn Stelbrink, Fadly Tantu, Hendrik Vogel, Kristina von Rintelen, Thomas von Rintelen, Katie Wagner, Margaret Whiting Blome, and Shane Wright each reviewed at least one chapter, and sometimes more. Anonymous referees offered comments on both the book proposal and draft book, leading to a host of revisions. I thank Lyubov Izmest'eva, Chris Martin, Greg Mayer, Matt McGee, and Alvaro Taveira for helpful discussions and exchanges. Randall Brummett, Ana Cristina Canales Gomez, Andy Cohen, William Darwall, Aneth David, Paola Ferreyros, Masanja Fortunatus, Juan Jose Miranda Montero, Michael Morris, Nare Ngoepe, Jusri Nilawati, Horace Owiti, Mike Pauers, Titus Bandulo Phiri, Walter Salzburger, Alena Shirokaya, Björn Stelbrink, Fadly Tantu, Martin Van der Knaap, and Griselle Felicita Vega helped identify quotations for the chapter openings. B. Blakely Brooks and Vladimir Condo generously aided with translation. Lyubov Izmest'eva, Tuma Kamulali, Axel Meyer, Mupape Mukuli, Lynne R. Parenti, and Björn Stelbrink kindly shared photographs. Oke Anakwenze and Brady Lloyd helped me keep working.

At the MIT Press, Anne-Marie Bono and Deborah Cantor-Adams were wonderful editors, and Debora Kuan and Oscar Sarkes were also a great help. Tim Lee helped with indexing. Haleigh Mooring persisted on the illustrations despite my many fussy requests. Steve Austad helped me begin, as did Larry Dill and Bambang Soeroto in different ways. Jim Costa and Steve Heard provided valuable guidance on the process. I am grateful for the hospitality of Doug Haffner, Peter Hehanussa, Axel Meyer, Jusri Nilawati, Walter Salzburger, Ole Seehausen, and Fadly Tantu during visits, and all that I learned. Doug Haffner helped me stick with ancient lakes.

I thank the University of Wisconsin–Whitewater, East Carolina University, Canadian International Development Agency, and US National Science Foundation for generous and essential support. I also thank the Furthermore program of the J. M. Kaplan fund.

The organizers of the Speciation/Species in Ancient Lakes conferences, a group of dedicated, enthusiastic, and generous researchers, have done science and nature a considerable service through their efforts. They introduced me to much of the work described here and several of the scientists. Some of the lead organizers (the list of all involved is much longer) of conferences I attended were Christian Albrecht, Andy Cohen, Doug Haffner, Ishmael Kimerei, Koen Martens, Axel Meyer, Frank Riedel, Casim Umba Tolo, Walter Salzburger, and Oleg Timoshkin.

Peter Hehanussa was a superb partner, friend, and ally. Dolph Schluter taught me a great deal and kindly shared his transporter illustration. Fadly Tantu and Jusri Nilawati became dear friends as well as valued colleagues during time spent together in Sulawesi (including their hospitality during a visit to Palu), North Carolina and Windsor; I remain grateful for all their generosity.

Glossary

Adaptive radiation The rapid diversification from an ancestor into various new species through natural selection arising from different environments or resources.

Allele One of the alternative forms of a gene.

Anthropocene A proposed geological epoch beginning with the start of significant human impact on the earth's ecosystems and geology.

Balancing selection Selection that favors more than one allele and maintains genetic variation in a population; includes negative frequency-dependent selection.

Convergent evolution The independent evolution of similar traits in separate lineages, usually in response to similar selection.

Disruptive selection Selection where extremes of a phenotypic distribution are favored over intermediate phenotypes within a population.

Divergent selection When different phenotypes are favored over intermediates in different environments (and usually populations).

Dominant allele An allele that produces the same phenotype whether paired with an identical or different allele.

Endemic A reference to a species confined to a limited area.

Environmental DNA Organismal DNA that can be found in the environment rather than being collected from an organism.

Frequency-dependent selection Selection where the fitness of an allele depends on its frequency in the population; selection favoring the rare allele is negative frequency dependent, and selection favoring the common allele is positive frequency dependent.

Genome The totality of an organism's DNA.

Heterozygous Having different alleles for the two copies of a gene (in a diploid organism like ourselves).

Homozygous Having the same allele for the two copies of a gene (in a diploid organism like ourselves).

Mutation Any change to the genomic sequence of an organism.

Natural selection A mechanism that can lead to adaptation whereby heritable differences in a trait cause some individuals to survive and reproduce more effectively than others.

Parallel evolution The independent evolution of similar traits, starting from a similar ancestral condition.

Phylogeny An evolutionary tree; a visual representation of the evolutionary history of species, populations, or genes.

Population A group of organisms belonging to the same species and living in the same area at the same time.

Purifying selection Selection that removes deleterious alleles from a population.

Recessive allele An allele that produces its characteristic features only when paired with an identical allele (i.e., homozygous).

Sexual selection Differential reproductive success resulting from competition for fertilization.

Speciation An evolutionary process by which new species arise, and one evolutionary lineage splits into two or more.

Species A set of populations (or members of a single large population) that interbreed and exchange genes only with each other (biological species concept).

Species flock A substantial number of closely related species confined to a limited area, such as a lake; closely related to adaptive radiation.

References and Further Reading

PREFACE

Clark, E. 1953. *Lady with a Spear*. New York: Harper.

Durrell, G. 1956. *My Family and Other Animals*. Harmondsworth, UK: Penguin.

Hampton, S. E., S. McGowan, T. Ozersky, S. G. P. Virdis, T. T. Vu, T. L. Spanbauer, B. M. Kraemer, et al. 2018. "Recent Ecological Change in Ancient Lakes." *Limnology and Oceanography* 63: 2277–2304.

Orwell, G. 1946. "Some Thoughts on the Common Toad." *Tribune*, April 12. Reprinted in *The Collected Essays, Journalism and Letters of George Orwell. Volume 4: In Front of Your Nose 1945–1950*. Harmondsworth, UK: Penguin.

Zim, H. S., R. W. Burnett, and H. L. Fisher. 1960. *Zoology: An Introduction to the Animal Kingdom*. Racine, WI: Western Publishing Company.

CHAPTER 1

Sulawesi and Telmatherinid Fish

Costa, J. T. 2014. *Wallace, Darwin and the Origin of Species*. Cambridge, MA: Harvard University Press.

Gray, S. M., and J. S. McKinnon. 2006. "A Comparative Description of Mating Behaviour in the Endemic Telmatherinid Fishes of Sulawesi's Malili Lakes." *Environmental Biology of Fishes* 75: 471–482.

Herder, F., A. W. Nolte, J. Pfaender, J. Schwarzer, R. K. Hadiaty, and U. K. Schliewen. 2006. "Adaptive Radiation and Hybridization in Wallace's

Dreamponds: Evidence from Sailfin Silversides in the Malili Lakes of Sulawesi." *Proceedings of the Royal Society B: Biological Sciences* 273: 2209–2217.

Kottelat, M. 1990. "Sailfin Silversides (Pisces: Telmatherinidae) of Lakes Towuti, Mahalona and Wawontoa (Sulawesi, Indonesia) with Descriptions of Two New Genera and Two New Species." *Ichthyological Exploration of Freshwaters* 1: 35–54.

Kottelat, M. 1991. "Sailfin Silversides (Pisces: Telmatherinidae) of Lake Matano, Sulawesi, Indonesia, with Descriptions of Six New Species." *Ichthyological Exploration of Freshwaters* 1: 321–344.

Wallace, A. R. 1869. *The Malay Archipelago: The Land of the Orang-Utan and the Bird of Paradise; a Narrative of Travel, with Studies of Man and Nature*. London: Macmillan.

Whitten, A. J., M. Mustafa, and G. S. Henderson. 1987. *The Ecology of Sulawesi.* Yogyakarta: Gadjah Mada University Press.

African Great Lakes and Their Cichlids

El Taher, A., F. Ronco, M. Matschiner, W. Salzburger, and A. Böhne. 2021. "Dynamics of Sex Chromosome Evolution in a Rapid Radiation of Cichlid Fishes." *Science Advances* 7: eabe8215.

Goldschmidt, T. 1996. *Darwin's Dreampond.* Cambridge, MA: MIT Press.

Marques, D. A., J. I. Meier, and O. Seehausen. 2019. "A Combinatorial View on Speciation and Adaptive Radiation." *Trends in Ecology & Evolution* 34: 531–544.

Meyer, A., Kocher, T. D., P. Basasibwaki, and A. C. Wilson. 1990. "Monophyletic Origin of Lake Victoria Cichlid Fishes Suggested by Mitochondrial DNA Sequences." *Nature* 347: 550–553.

Miller, E. C. 2021. "Comparing Diversification Rates in Lakes, Rivers, and the Sea." *Evolution* 75: 2055–2073.

Roberts, R. B., J. R. Ser, and T. D. Kocher. 2009. "Sexual Conflict Resolved by Invasion of a Novel Sex Determiner in Lake Malawi Cichlid Fishes." *Science* 326: 998–1001.

Salzburger, W., B. Van Bocxlaer, and A. S. Cohen. 2014. "Ecology and Evolution of the African Great Lakes and Their Faunas." *Annual Review of Ecology, Evolution, and Systematics* 45: 519–545.

Taborsky, M. 2016. "Cichlid Fishes: A Model for the Integrative Study of Social Behavior." In *Cooperative Breeding in Vertebrates*, edited by W. D. Koenig and J. L. Dickinson, 272–293. Cambridge: Cambridge University Press.

Ancient Lake Overviews

Brooks, J. L. 1950. "Speciation in Ancient Lakes." *Quarterly Review of Biology* 25: 30–60.

Couston, L. A., and M. Siegert. 2021. "Dynamic Flows Create Potentially Habitable Conditions in Antarctic Subglacial Lakes." *Science Advances* 7: eabc3972.

Cristescu, M. E., S. J. Adamowicz, J. J. Vaillant, and D. G. Haffner. 2010. "Ancient Lakes Revisited: From the Ecology to the Genetics of Speciation." *Molecular Ecology* 19: 4837–4851.

Hampton, S. E., S. McGowan, T. Ozersky, S. G. P. Virdis, T. T. Vu, T. L. Spanbauer, B. M. Kraemer, et al. 2018. "Recent Ecological Change in Ancient Lakes." *Limnology and Oceanography* 63: 2277–2304.

Kawanabe, H., G. W. Coulter, and A. C. Roosevelt. 1999. *Ancient Lakes: Their Cultural and Biological Diversity*. Ghent: Kenobi Productions.

Martens, K. 1997. "Speciation in Ancient Lakes." *Trends in Ecology & Evolution* 12: 177–182.

Rossiter, A., and H. Kawanabe. 2000. *Ancient Lakes: Biodiversity, Ecology and Evolution*. London: Academic Press.

Baikal

Hampton, S. E., R. Lyubov, M. V. Izmest'eva, S. L. Moore, B. D. Katz, and E. A. Silow. 2008. "Sixty Years of Environmental Change in the World's Largest Freshwater Lake–Lake Baikal, Siberia." *Global Change Biology* 14: 1947–1958.

Macdonald, K. S., III, L. Yampolsky, and J. E. Duffy. 2005. "Molecular and Morphological Evolution of the Amphipod Radiation of Lake Baikal." *Molecular Phylogenetics and Evolution* 35: 323–343.

Sherbakov, D. Y. 1999. "Molecular Phylogenetic Studies on the Origin of Biodiversity in Lake Baikal." *Trends in Ecology & Evolution* 14: 92–95.

Thomson, P. 2007. *Sacred Sea: A Journey to Lake Baikal*. Oxford: Oxford University Press.

Lake Sediments

Cohen, A. S. 2012. "Scientific Drilling and Biological Evolution in Ancient Lakes: Lessons Learned and Recommendations for the Future." *Hydrobiologia* 682: 3–25.

Cohen, A. S. 2018. "The Past Is a Key to the Future: Lessons Paleoecological Data Can Provide for Management of the African Great Lakes." *Journal of Great Lakes Research* 44: 1142–1153.

Wilke, T., T. Hauffe, E. Jovanovska, A. Cvetkoska, T. Donders, K. Ekschmitt, A. Francke, et al. 2020. "Deep Drilling Reveals Massive Shifts in Evolutionary Dynamics after Formation of Ancient Ecosystem." *Science Advances* 6: eabb2943.

Wilke, T., B. Wagner, B. Van Bocxlaer, C. Albrecht, D. Ariztegui, D. Delicado, A. Francke, et al. 2016. "Scientific Drilling Projects in Ancient Lakes: Integrating Geological and Biological Histories." *Global and Planetary Change* 143: 118–151.

CHAPTER 2

Henderson, H. G. 1958. *An Introduction to Haiku: An Anthology of Poems and Poets from Basho to Shiki*. Garden City, NY: Doubleday Anchor.

Species Concepts

Bobay, L. M., and H. Ochman. 2018. Biological Species in the Viral World. *Proceedings of the National Academy of Sciences* 115: 6040–6045.

Coyne, J. A. 2005. "Ernst Mayr (1904–2005)." *Science* 307: 1212–1213.

Coyne, J. A., and H. A. Orr. 2004. *Speciation*. Sunderland, MA: Sinauer Associates.

Darwin, C. 1859. *On the Origin of Species by Means of Natural Selection*. London: John Murray.

Mandagi, I. F., R. Kakioka, J. Montenegro, H. Kobayashi, K. W. Masengi, N. Inomata, A. J. Nagano, et al. 2021. "Species Divergence and Repeated Ancient Hybridization in a Sulawesian Lake System." *Journal of Evolutionary Biology* 34: 1767–1780.

Stankowski, S., and M. Ravinet. 2021. Defining the Speciation Continuum. *Evolution* 75: 1256–1273.

Stankowski, S. and M. Ravinet. 2021. Quantifying the Use of Species Concepts. *Current Biology*, 31: R428–R429.

Measuring Biodiversity

Herder, F., A. W. Nolte, J. Pfaender, J. Schwarzer, R. K. Hadiaty, and U. K. Schliewen. 2006. "Adaptive Radiation and Hybridization in Wallace's Dreamponds: Evidence from Sailfin Silversides in the Malili Lakes of Sulawesi." *Proceedings of the Royal Society B: Biological Sciences* 273: 2209–2217.

Mandagi, I. F., R. Kakioka, J. Montenegro, H. Kobayashi, K. W. Masengi, N. Inomata, A. J. Nagano, et al. 2021. "Species Divergence and Repeated Ancient Hybridization in a Sulawesian Lake System." *Journal of Evolutionary Biology* 34: 1767–1780.

Miesen, F. W., F. Droppelmann, S. Hullen, R. K. Hadiaty, and F. Herder. 2018. "An Annotated Checklist of the Inland Fishes of Sulawesi." *Bonn Zoological Bulletin* 64: 77–106.

Redding, D. W., and A. Ø. Mooers. 2006. "Incorporating Evolutionary Measures into Conservation Prioritization." *Conservation Biology* 20: 1670–1678.

Veron, S., V. Saito, N. Padilla-García, F. Forest, and Y. Bertheau. 2019. "The Use of Phylogenetic Diversity in Conservation Biology and Community Ecology: A Common Base but Different Approaches." *Quarterly Review of Biology* 94: 123–148.

Speciation (See Also "Species Concepts" Above)

Herder, F., J. Pfaender, and U. K. Schliewen. 2008. "Adaptive Sympatric Speciation of Polychromatic 'Roundfin' Sailfin Silverside Fish in Lake Matano (Sulawesi)." *Evolution* 62: 2178–2195.

Lewontin, R. C. 1970. "The Units of Selection." *Annual Review of Ecology and Systematics* 1: 1–18.

Meier, J. I., D. A. Marques, C. E. Wagner, L. Excoffier, and O. Seehausen. 2018. "Genomics of Parallel Ecological Speciation in Lake Victoria Cichlids." *Molecular Biology and Evolution* 35: 1489–1506.

Meier, J. I., V. C. Sousa, D. A. Marques, O. M. Selz, C. E. Wagner, L. Excoffier, and O. Seehausen. 2017. "Demographic Modelling with Whole-Genome Data Reveals Parallel Origin of Similar *Pundamilia* Cichlid Species after Hybridization." *Molecular Ecology* 26: 123–141.

Nosil, P. 2012. *Ecological Speciation*. Oxford: Oxford University Press.

Rajkov, J., A. A. Weber, W. Salzburger, and B. Egger. 2018. "Immigrant and Extrinsic Hybrid Inviability Contribute to Reproductive Isolation between Lake and River Cichlid Ecotypes." *Evolution* 72: 2553–2564.

Schluter, D. 2009. "Evidence for Ecological Speciation and Its Alternative." *Science* 323: 737–741.

Sutra, N., J. Kusumi, J. Montenegro, H. Kobayashi, S. Fujimoto, K. W. A. Masengi, A. J. Nagano, et al. 2019. "Evidence for Sympatric Speciation in a Wallacean Ancient Lake." *Evolution* 73: 1898–1915.

Adaptive Radiation

Albertson, R. C., and T. D. Kocher. 2006. "Genetic and Developmental Basis of Cichlid Trophic Diversity." *Heredity* 97: 211–221.

Chaparro-Pedraza, P. C., G. Roth, and O. Seehausen. 2022. "The Enrichment Paradox in Adaptive Radiations: Emergence of Predators Hinders Diversification in Resource Rich Environments." *Ecology Letters*, 1–12.

Gillespie, R. G., G. M. Bennett, L. De Meester, J. L. Feder, R. C. Fleischer, L. J. Harmon, A. P. Hendry, et al. 2020. "Comparing Adaptive Radiations across Space, Time, and Taxa." *Journal of Heredity* 111: 1–20.

Kocher, T. D., J. A. Conroy, K. R. McKaye, and J. R. Stauffer. 1993. "Similar Morphologies of Cichlid Fish in Lakes Tanganyika and Malawi Are Due to Convergence." *Molecular Phylogenetics and Evolution* 2: 158–165.

McGee, M. D., S. R. Borstein, J. I. Meier, D. A. Marques, S. Mwaiko, A. Taabu, M. A. Kishe, et al. 2020. "The Ecological and Genomic Basis of Explosive Adaptive Radiation." *Nature* 586: 75–79.

Miura, O., M. Urabe, T. Nishimura, K. Nakai, and S. Chiba. 2019. "Recent Lake Expansion Triggered the Adaptive Radiation of Freshwater Snails in the Ancient Lake Biwa." *Evolution Letters* 3: 43–54.

Muschick, M., A. Indermaur, and W. Salzburger. 2012. "Convergent Evolution within an Adaptive Radiation of Cichlid Fishes." *Current Biology* 22: 2362–2368.

Parenti, L. R., and D. Wowor. 2020. "Renny Kurnia Hadiaty (1960–2019)." *Copeia* 108: 430–433.

Pfaender, J., R. K. Hadiaty, U. K. Schliewen, and F. Herder. 2016. "Rugged Adaptive Landscapes Shape a Complex, Sympatric Radiation." *Proceedings of the Royal Society B: Biological Sciences* 283: 20152342.

Schluter, D. 2000. *The Ecology of Adaptive Radiation*. Oxford: Oxford University Press.

Tabata, R., R. Kakioka, K. Tominaga, T. Komiya, and K. Watanabe. 2016. "Phylogeny and Historical Demography of Endemic Fishes in Lake Biwa: The Ancient Lake as a Promoter of Evolution and Diversification of Freshwater Fishes in Western Japan." *Ecology and Evolution* 6: 2601–2623.

Wagner, C. E., L. J. Harmon, and O. Seehausen. 2012. "Ecological Opportunity and Sexual Selection Together Predict Adaptive Radiation." *Nature* 487: 366–369.

Wagner, C. E., L. J. Harmon, and O. Seehausen. 2014. "Cichlid Species-Area Relationships Are Shaped by Adaptive Radiations That Scale with Area." *Ecology Letters* 17: 583–592.

Wilson, A. B., M. Glaubrecht, and A. Meyer. 2004. "Ancient Lakes as Evolutionary Reservoirs: Evidence from the Thalassoid Gastropods of Lake Tanganyika." *Proceedings of the Royal Society B: Biological Sciences* 271: 529–536.

CHAPTER 3

Woodward, S. P. 1859. "On Some New Freshwater Shells from Central Africa." *Proceedings of the Zoological Society of London* 27: 348–350.

Snails and Crabs

Glaubrecht, M., and T. von Rintelen. 2008. "The Species Flocks of Lacustrine Gastropods: *Tylomelania* on Sulawesi as Models in Speciation and Adaptive Radiation." In *Patterns and Processes of Speciation in Ancient Lakes*, edited by T. Wilke, R. Väinölä, and F. Riedel, 181–199. Dordrecht: Springer.

Hilgers, L., S. Hartmann, M. Hofreiter, and T. von Rintelen. 2018. "Novel Genes, Ancient Genes, and Gene Co-option Contributed to the Genetic Basis of the Radula, a Molluscan Innovation." *Molecular Biology and Evolution* 35: 1638–1652.

Hilgers, L., S. Hartmann, J. Pfaender, N. Lentge-Maaß, R. M. Marwoto, T. von Rintelen, and M. Hofreiter. 2022. "Radula Diversification Promotes Ecomorph Divergence in an Adaptive Radiation of Freshwater Snails." *Genes* 13: 1029. https://doi.org/10.3390/genes13061029.

Poettinger, T., and C. D. Schubart. 2014. "Molecular Diversity of Freshwater Crabs from Sulawesi and the Sequential Colonization of Ancient Lakes." *Hydrobiologia* 739: 73–84.

Schubart, C. D., and P. K. L. Ng. 2008. "A New Molluscivore Crab from Lake Poso Confirms Multiple Colonization of Ancient Lakes in Sulawesi by Freshwater

Crabs (Decapoda: Brachyura)." *Zoological Journal of the Linnean Society* 154: 211–221.

von Rintelen, T., P. Bouchet, and M. Glaubrecht. 2007. "Ancient Lakes as Hotspots of Diversity: A Morphological Review of an Endemic Species Flock of *Tylomelania* (Gastropoda: Cerithioidea: Pachychilidae) in the Malili Lake System on Sulawesi, Indonesia." *Hydrobiologia* 592: 11–94.

von Rintelen, T., and M. Glaubrecht. 2005. "Anatomy of an Adaptive Radiation: A Unique Reproductive Strategy in the Endemic Freshwater Gastropod *Tylomelania* (Cerithioidea: Pachychilidae) on Sulawesi, Indonesia and Its Biogeographical Implications." *Biological Journal of the Linnean Society* 85: 513–542.

von Rintelen, T., A. B. Wilson, A. Meyer, and M. Glaubrecht. 2004. "Escalation and Trophic Specialization Drive Adaptive Radiation of Freshwater Gastropods in Ancient Lakes on Sulawesi, Indonesia." *Proceedings of the Royal Society B: Biological Sciences* 271: 2541–2549.

West, K., and A. Cohen. 1994. "Predator-Prey Coevolution as a Model for the Unusual Morphologies of the Crabs and Gastropods of Lake Tanganyika." *Ergebnisse der Limnologie* 44: 267–283.

West, K., A. Cohen, and M. Baron. 1991. "Morphology and Behavior of Crabs and Gastropods from Lake Tanganyika, Africa: Implications for Lacustrine Predator-Prey Coevolution." *Evolution* 45: 589–607.

Wilson, A. B., M. Glaubrecht, and A. Meyer. 2004. "Ancient Lakes as Evolutionary Reservoirs: Evidence from the Thalassoid Gastropods of Lake Tanganyika." *Proceedings of the Royal Society B: Biological Sciences* 271: 529–536.

Sponges

Erpenbeck, D., A. Galitz, G. Wörheide, C. Albrecht, R. Pronzato, and R. Manconi. 2020. "Having the Balls to Colonize: The *Ephydatia fluviatilis* Group and the Origin of (Ancient) Lake 'Endemic' Sponge Lineages." *Journal of Great Lakes Research* 46: 1140–1145.

Kenny, N. J., W. R. Francis, R. E. Rivera-Vicéns, K. Juravel, A. de Mendoza, C. Díez-Vives, R. Lister, et al. 2020. "Tracing Animal Genomic Evolution with the Chromosomal-Level Assembly of the Freshwater Sponge *Ephydatia muelleri*." *Nature Communications* 11: 1–11.

Kenny, N. J., and V. B. Itskovich. 2020. "Phylogenomic Inference of the Interrelationships of Lake Baikal Sponges." *Systematics and Biodiversity* 19: 209–217.

Kenny, N. J., B. Plese, A. Riesgo, and V. B. Itskovich. 2019. "Symbiosis, Selection, and Novelty: Freshwater Adaptation in the Unique Sponges of Lake Baikal." *Molecular Biology and Evolution* 36: 2462–2480.

Manconi, R., and R. Pronzato. 2016. "How to Survive and Persist in Temporary Freshwater? Adaptive Traits of Sponges (Porifera: Spongillida): A Review." *Hydrobiologia* 782: 11–22.

Van Soest, R. W. M., N. Boury-Esnault, J. Vacelet, M. Dohrmann, D. Erpenbeck, N. J. De Voogd, N. Santodomingo, et al. 2012. "Global Diversity of Sponges (Porifera)." *PLoS ONE* 7: e35105.

Shrimp (Also Cichlid Dispersal)

Brooks, K. C., R. Maia, J. E. Duffy, K. M. Hultgren, and D. R. Rubenstein. 2017. "Ecological Generalism Facilitates the Evolution of Sociality in Snapping Shrimps." *Ecology Letters* 20: 1516–1525.

Horká, I., S. De Grave, C. H. J. M. Fransen, A. Petrusek, and Z. Ďuriš. 2016. "Multiple Host Switching Events Shape the Evolution of Symbiotic Palaemonid Shrimps (Crustacea: Decapoda)." *Scientific Reports* 6: 1–13.

Husemann, M., M. Tobler, B. Ding, R. Nguyen, C. McCauley, T. Pilger, and P. D. Danley. 2019. "Complex Patterns of Genetic and Phenotypic Divergence in Populations of the Lake Malawi Cichlid *Maylandia zebra*." *Hydrobiologia* 832: 135–151.

Klotz, W., T. von Rintelen, D. Wowor, C. Lukhaup, and K. von Rintelen. 2021. "Lake Poso's Shrimp Fauna Revisited: The Description of Five New Species of the Genus *Caridina* (Crustacea, Decapoda, Atyidae) More than Doubles the Number of Endemic Lacustrine Species." *ZooKeys* 1009: 81–122.

Koblmüller, S., W. Salzburger, B. Obermüller, E. V. A. Eigner, C. Sturmbauer, and K. M. Sefc. 2011. "Separated by Sand, Fused by Dropping Water: Habitat Barriers and Fluctuating Water Levels Steer the Evolution of Rock-Dwelling Cichlid Populations in Lake Tanganyika." *Molecular Ecology* 20: 2272–2290.

von Rintelen, K. 2011. "Intraspecific Geographic Differentiation and Patterns of Endemism in Freshwater Shrimp Species Flocks in Ancient Lakes of Sulawesi." In *Phylogeography and Population Genetics in Crustacea*, edited by C. Held, S. Koenemann, and C. D. Schubart, 257–271. Boca Raton: CRC Press.

von Rintelen, K., and Y. Cai. 2009. "Radiation of Endemic Species Flocks in Ancient Lakes: Systematic Revision of the Freshwater Shrimp *Caridina* H.

Milne Edwards, 1837 (Crustacea: Decapoda: Atyidae) from the Ancient Lakes of Sulawesi, Indonesia, with the Description of Eight New Species." *Raffles Bulletin of Zoology* 57: 343–452.

von Rintelen, K., P. De los Ríos, T. von Rintelen, G. C. B. Poore, and M. Thiel. 2020. "Standing Waters, Especially Ancient Lakes." In *Evolution and Biogeography*, vol. 8, edited by M. Thiel and G. Poore, 296–318. New York: Oxford University Press.

von Rintelen, K., M. Glaubrecht, C. D. Schubart, A. Wessel, and T. von Rintelen. 2010. "Adaptive Radiation and Ecological Diversification of Sulawesi's Ancient Lake Shrimps." *Evolution* 64: 3287–3299.

von Rintelen, K., T. von Rintelen, M. Meixner, C. Lüter, Y. Cai, and M. Glaubrecht. 2007. "Freshwater Shrimp–Sponge Association from an Ancient Lake." *Biology Letters* 3: 262–264.

Zitzler, K., and Y. Cai. 2006. "*Caridina spongicola*, New Species, a Freshwater Shrimp (Crustacea: Decapoda: Atyidae) from the Ancient Malili Lake System of Sulawesi, Indonesia." *Raffles Bulletin of Zoology* 54: 271–276.

Cichlids of Lake Palaeo-Makgadikgadi

Astudillo-Clavijo, V., M. L. Stiassny, K. L. Ilves, Z. Musilova, W. Salzburger, and H. López-Fernández. 2022. "Exon-Based Phylogenomics and the Relationships of African Cichlid Fishes: Tackling the Challenges of Reconstructing Phylogenies with Repeated Rapid Radiations." *Systematic Biology* syac051.

Joyce, D. A., D. H. Lunt, R. Bills, G. F. Turner, C. Katongo, N. Duftner, C. Sturmbauer, and O. Seehausen. 2005. "An Extant Cichlid Fish Radiation Emerged in an Extinct Pleistocene Lake." *Nature* 435: 90–95.

Schmidt, M., M. Fuchs, A. C. G. Henderson, A. Kossler, M. J. Leng, A. W. Mackay, E. Shemang, and F. Riedel. 2017. "Paleolimnological Features of a Mega-Lake Phase in the Makgadikgadi Basin (Kalahari, Botswana) during Marine Isotope Stage 5 Inferred from Diatoms." *Journal of Paleolimnology* 58: 373–390.

CHAPTER 4

Mwenelupembe, P. 2013. "Lake Malawi" (poem). Accessed September 22, 2022. https://www.poemhunter.com/poem/lake-malawi.

Vision and Light

Endler, J. A. 1990. "On the Measurement and Classification of Colour in Studies of Animal Colour Patterns." *Biological Journal of the Linnean Society* 41: 315–352.

McGee, M. D., S. R. Borstein, J. I. Meier, D. A. Marques, S. Mwaiko, A. Taabu, M. A. Kishe, et al. 2020. "The Ecological and Genomic Basis of Explosive Adaptive Radiation." *Nature* 586: 75–79.

Seehausen, O., and J. J. M. van Alphen. 1998. "The Effect of Male Coloration on Female Mate Choice in Closely Related Lake Victoria Cichlids (*Haplochromis nyererei* Complex)." *Behavioral Ecology and Sociobiology* 42: 1–8.

Seehausen, O., J. J. M. van Alphen, and F. Witte. 1997. "Cichlid Fish Diversity Threatened by Eutrophication That Curbs Sexual Selection." *Science* 277: 1808–1811.

Selz, O. M., M. E. R. Pierotti, M. E. Maan, C. Schmid, and O. Seehausen. 2014. "Female Preference for Male Color is Necessary and Sufficient for Assortative Mating in 2 Cichlid Sister Species." *Behavioral Ecology* 25: 612–626.

Van der Sluijs, I., S. M. Gray, M. C. P. Amorim, I. Barber, U. Candolin, A. P. Hendry, R. Krahe, et al. 2011. "Communication in Troubled Waters: Responses of Fish Communication Systems to Changing Environments." *Evolutionary Ecology* 25: 623–640.

Wagner, C. E., L. J. Harmon, and O. Seehausen. 2012. "Ecological Opportunity and Sexual Selection Together Predict Adaptive Radiation." *Nature* 487: 366–369.

Wagner, C. E., L. J. Harmon, and O. Seehausen. 2014. "Cichlid Species-Area Relationships Are Shaped by Adaptive Radiations That Scale with Area." *Ecology Letters* 17: 583–592.

Competing for Mates: Sexual Selection

Darwin, C. 1871. *The Descent of Man, and Selection in Relation to Sex*. London: Murray.

Kokko, H., H. Klug, and M. D. Jennions. 2012. "Unifying Cornerstones of Sexual Selection: Operational Sex Ratio, Bateman Gradient and the Scope for Competitive Investment." *Ecology Letters* 15: 1340–1351.

Pauers, M. J., J. S. McKinnon, and T. J. Ehlinger. 2004. "Directional Sexual Selection on Chroma and Within: Pattern Colour Contrast in *Labeotropheus fuelleborni*." *Proceedings of the Royal Society B: Biological Sciences* 271: S444–S447.

Trivers, R. 1972. "Parental Investment and Sexual Selection." In *Sexual Selection and the Descent of Man: The Darwinian Pivot*, edited by B. Campbell, 136–179. Chicago: Aldine Press.

The Runaway Hypothesis of Sexual Selection

Fisher, R. A. 1930. *The Genetical Theory of Natural Selection*. Oxford: Clarendon Press.

Lande, R. 1981. "Models of Speciation by Sexual Selection on Polygenic Traits." *Proceedings of the National Academy of Sciences* 78: 3721–3725.

Paul, D. B., and H. G. Spencer. 1995. "The Hidden Science of Eugenics." *Nature* 374: 302–304.

Reilly, P. R. 2015. "Eugenics and Involuntary Sterilization: 1907–2015." *Annual Review of Genomics and Human Genetics* 16: 351–368.

Servedio, M. R., and J. W. Boughman. 2017. "The Role of Sexual Selection in Local Adaptation and Speciation." *Annual Review of Ecology, Evolution, and Systematics* 48: 85–109.

Sensory Drive

Cummings, M. E., and J. A. Endler. 2018. "25 Years of Sensory Drive: The Evidence and Its Watery Bias." *Current Zoology* 64: 471–484.

Fuller, R. C., and J. A. Endler. 2018. "A Perspective on Sensory Drive." *Current Zoology* 64: 465–470.

Gray, S. M., L. M. Dill, F. Y. Tantu, E. R. Loew, F. Herder, and J. S. McKinnon. 2008. "Environment-Contingent Sexual Selection in a Colour Polymorphic Fish." *Proceedings of the Royal Society B: Biological Sciences* 275: 1785–1791.

Gray, S. M., and J. S. McKinnon. 2007. "Linking Color Polymorphism Maintenance and Speciation." *Trends in Ecology & Evolution* 22: 71–79.

Maan, M. E., O. Seehausen, and T. G. G. Groothuis. 2017. "Differential Survival between Visual Environments Supports a Role of Divergent Sensory Drive in Cichlid Fish Speciation." *American Naturalist* 189: 78–85.

Seehausen, O., Y. Terai, I. S. Magalhaes, K. L. Carleton, H. D. J. Mrosso, R. Miyagi, I. Van Der Sluijs, et al. 2008. "Speciation through Sensory Drive in Cichlid Fish." *Nature* 455: 620–626.

Learning and Development

Verzijden, M. N., and C. ten Cate. 2007. "Early Learning Influences Species Assortative Mating Preferences in Lake Victoria Cichlid Fish." *Biology Letters* 3: 134–136.

Verzijden, M. N., C. ten Cate, M. R. Servedio, G. M. Kozak, J. W. Boughman, and E. I. Svensson. 2012. "The Impact of Learning on Sexual Selection and Speciation." *Trends in Ecology & Evolution* 27: 511–519.

Wright, D. S., N. Demandt, J. T. Alkema, O. Seehausen, T. G. G. Groothuis, and M. E. Maan. 2017. "Developmental Effects of Visual Environment on Species-Assortative Mating Preferences in Lake Victoria Cichlid Fish." *Journal of Evolutionary Biology* 30: 289–299.

Wright, D. S., E. Rietveld, and M. E. Maan 2018. "Developmental Effects of Environmental Light on Male Nuptial Coloration in Lake Victoria Cichlid Fish." *PeerJ* 6: e4209.

Wright, D. S., R. van Eijk, L. Schuart, O. Seehausen, T. G. G. Groothuis, and M. E. Maan. 2020. "Testing Sensory Drive Speciation in Cichlid Fish: Linking Light Conditions to Opsin Expression, Opsin Genotype and Female Mate Preference." *Journal of Evolutionary Biology* 33: 422–434.

Conflict

Arnqvist, G., and L. Rowe. 2005. *Sexual Conflict.* Princeton, NJ: Princeton University Press.

Gray, S. M., L. M. Dill, and J. S. McKinnon. 2007. "Cuckoldry Incites Cannibalism: Male Fish Turn to Cannibalism When Perceived Certainty of Paternity Decreases." *American Naturalist* 169: 258–263.

Gray, S. M., and J. S. McKinnon. 2006. "A Comparative Description of Mating Behaviour in the Endemic Telmatherinid Fishes of Sulawesi's Malili Lakes." *Environmental Biology of Fishes* 75: 471–482.

Kocher, T. D., K. A. Behrens, M. A. Conte, M. Aibara, H. D. J. Mrosso, E. C. J. Green, M. R. Kidd, et al. 2022. "New Sex Chromosomes in Lake Victoria Cichlid Fishes (Cichlidae: Haplochromini)." *Genes* 13: 804.

Moore, E. C., P. J. Ciccotto, E. N. Peterson, M. S. Lamm, R. C. Albertson, and R. B. Roberts. 2022. "Polygenic Sex Determination Produces Modular Sex Polymorphism in an African Cichlid Fish." *Proceedings of the National Academy of Sciences* 119: e2118574119.

Rometsch, S. J., J. Torres-Dowdall, and A. Meyer. 2020. "Evolutionary Dynamics of Pre- and Postzygotic Reproductive Isolation in Cichlid Fishes." *Philosophical Transactions of the Royal Society B* 375: 20190535.

CHAPTER 5

Herder, F., A. W. Nolte, J. Pfaender, J. Schwarzer, R. K. Hadiaty, and U. K. Schliewen. 2006. "Adaptive Radiation and Hybridization in Wallace's Dreamponds: Evidence from Sailfin Silversides in the Malili Lakes of Sulawesi." *Proceedings of the Royal Society B: Biological Sciences* 273: 2209–2217.

Lefty Sporting Advantage and Balancing Selection

Brooks, R., L. F. Bussiere, M. D. Jennions, and J. Hunt. 2004. "Sinister Strategies Succeed at the Cricket World Cup." *Proceedings of the Royal Society B: Biological Sciences* 271: S64–S66.

Jamie, G. A., and J. I. Meier. 2020. "The Persistence of Polymorphisms across Species Radiations." *Trends in Ecology & Evolution* 35: 795–808.

Raymond, M., D. Pontier, A. Dufour, and A. P. Møller. 1996. "Frequency-Dependent Maintenance of Left Handedness in Humans." *Proceedings of the Royal Society B: Biological Sciences* 263: 1627–1633.

Scale Eating

Hori, M. 1993. "Frequency-Dependent Natural Selection in the Handedness of Scale-Eating Cichlid Fish." *Science* 260: 216–219.

Indermaur, A., A. Theis, B. Egger, and W. Salzburger. 2018. Mouth Dimorphism in Scale-Eating Cichlid Fish from Lake Tanganyika Advances Individual Fitness." *Evolution* 72: 1962–1969.

Kusche, H., H. J. Lee, and A. Meyer. 2012. "Mouth Asymmetry in the Textbook Example of Scale-Eating Cichlid Fish Is Not a Discrete Dimorphism after All." *Proceedings of the Royal Society B: Biological Sciences* 279: 4715–4723.

Lee, H. J., V. Heim, and A. Meyer. 2015. "Genetic and Environmental Effects on the Morphological Asymmetry in the Scale-Eating Cichlid Fish, *Perissodus microlepis*." *Ecology and Evolution* 5: 4277–4286.

Lee, H. J., H. Kusche, and A. Meyer. 2012. "Handed Foraging Behavior in Scale-Eating Cichlid Fish: Its Potential Role in Shaping Morphological Asymmetry." *PLoS ONE* 7: e44670.

Liem, K. F., and D. J. Stewart. 1976. "Evolution of the Scale-Eating Cichlid Fishes of Lake Tanganyika: A Generic Revision with a Description of a New Species." *Bulletin of the Museum of Comparative Zoology* 147: 1975–1977.

Marlier, G., and N. Leleup. 1954. "A Curious Ecological 'Niche' among the Fishes of Lake Tanganyika." *Nature* 174: 935–936.

Palmer, A. R. 2009. "Animal Asymmetry." *Current Biology* 19: R473–R477.

Palmer, A. R. 2010. "Scale-Eating Cichlids: From Hand(ed) to Mouth." *Journal of Biology* 9: 1–4.

Raffini, F., and A. Meyer. 2019. "A Comprehensive Overview of the Developmental Basis and Adaptive Significance of a Textbook Polymorphism: Head Asymmetry in the Cichlid Fish *Perissodus microlepis*." *Hydrobiologia* 832: 65–84.

Takeuchi, Y., M. Hori, S. Tada, and Y. Oda. 2016. "Acquisition of Lateralized Predation Behavior Associated with Development of Mouth Asymmetry in a Lake Tanganyika Scale-Eating Cichlid Fish." *PLoS ONE* 11: e0147476.

Takeuchi, Y., and Y. Oda. 2017. "Lateralized Scale-Eating Behaviour of Cichlid Is Acquired by Learning to Use the Naturally Stronger Side." *Scientific Reports*, 1–9.

Van Dooren, T. J. M., H. A. Van Goor, and M. Van Putten. 2010. "Handedness and Asymmetry in Scale-Eating Cichlids: Antisymmetries of Different Strength." *Evolution* 64: 2159–2165.

Sexual Trickery

Cerwenka, A. F., J. D. Wedekind, R. K. Hadiaty, U. K. Schliewen, and F. Herder. 2012. "Alternative Egg-Feeding Tactics in *Telmatherina sarasinorum*, a Trophic Specialist of Lake Matano's Evolving Sailfin Silversides Fish Radiation." *Hydrobiologia* 693: 131–139.

Gray, S. M., and J. S. McKinnon. 2006. A Comparative Description of Mating Behaviour in the Endemic Telmatherinid Fishes of Sulawesi's Malili Lakes." *Environmental Biology of Fishes* 75: 471–482.

Gray, S. M., J. S. McKinnon, F. Y. Tantu, and L. M. Dill. 2008. "Sneaky Egg-Eating in *Telmatherina sarasinorum*, an Endemic Fish from Sulawesi." *Journal of Fish Biology* 73: 728–731.

Lloyd, James E. 1965. "Aggressive Mimicry in *Photuris*: Firefly Femmes Fatales." *Science* 149: 653–654.

Color, Contrast, Courtship, and Conflict

Chunco, A. J., J. S. McKinnon, and M. R. Servedio. 2007. "Microhabitat Variation and Sexual Selection Can Maintain Male Color Polymorphisms." *Evolution* 61: 2504–2515.

Dijkstra, P. D., and S. E. Border. 2018. "How Does Male–Male Competition Generate Negative Frequency-Dependent Selection and Disruptive Selection during Speciation?" *Current Zoology* 64: 89–99.

Seehausen, O., and D. Schluter. 2004. "Male–Male Competition and Nuptial-colour Displacement as a Diversifying Force in Lake Victoria Cichlid Fishes." *Proceedings of the Royal Society B: Biological Sciences* 271: 1345–1353.

CHAPTER 6

Hybridization in Neanderthals and Fish

Ballard, J. W. O., and M. C. Whitlock. 2004. "The Incomplete Natural History of Mitochondria." *Molecular Ecology* 13: 729–744.

Blackwell, T., A. G. Ford, A .G. Ciezarek, S. J. Bradbeer, C. A. Gracida Juarez, A. M. Smith, B. P. Ngatunga, et al. 2021. "Newly Discovered Cichlid Fish Biodiversity Threatened by Hybridization with Non-Native Species." *Molecular Ecology* 30: 895–911.

Herder, F., A. W. Nolte, J. Pfaender, J. Schwarzer, R. K. Hadiaty, and U. K. Schliewen. 2006. "Adaptive Radiation and Hybridization in Wallace's Dreamponds: Evidence from Sailfin Silversides in the Malili Lakes of Sulawesi." *Proceedings of the Royal Society B: Biological Sciences* 273: 2209–2217

Keller, I., C. E. Wagner, L. Greuter, S. Mwaiko, O. M. Selz, A. Sivasundar, S. Wittwer, and O. Seehausen. 2013. "Population Genomic Signatures of Divergent Adaptation, Gene Flow and Hybrid Speciation in the Rapid Radiation of Lake Victoria Cichlid Fishes." *Molecular Ecology* 22: 2848–2863.

Shechonge, A., B. P. Ngatunga, S. J. Bradbeer, J. J. Day, J. J. Freer, A. G. P. Ford, J. Kihedu, et al. 2019. "Widespread Colonisation of Tanzanian Catchments by Introduced *Oreochromis* Tilapia Fishes: The Legacy from Decades of Deliberate Introduction." *Hydrobiologia* 832: 235–253.

Vijaykrishna, D., L. L. M. Poon, H. C. Zhu, S. K. Ma, O. T. W. Li, C. L. Cheung, G. J. D. Smith, et al. 2010. "Reassortment of Pandemic H1N1/2009 Influenza A Virus in Swine." *Science* 328: 1529–1529.

Zeberg, H., and S. Pääbo. 2020. "The Major Genetic Risk Factor for Severe COVID-19 Is Inherited from Neanderthals." *Nature*, 610–612.

Zeberg, H., and S. Pääbo. 2021. "A Genomic Region Associated with Protection against Severe COVID-19 Is Inherited from Neandertals." *Proceedings of the National Academy of Sciences* 118: e2026309118.

The Transgressions of Genes

Lewontin, R. C., and L. C. Birch. 1966. "Hybridization as a Source of Variation for Adaptation to New Environments." *Evolution* 20: 315–336.

Richards, E. J., and C. H. Martin. 2017. "Adaptive Introgression from Distant Caribbean Islands Contributed to the Diversification of a Microendemic Adaptive Radiation of Trophic Specialist Pupfishes." *PLoS Genetics* 13: e1006919.

Richards, E. J., and C. H. Martin. 2022. "We Get By with a Little Help from Our Friends: Shared Adaptive Variation Provides a Bridge to Novel Ecological Specialists during Adaptive Radiation." *Proceedings of the Royal Society B* 289: 20220613.

Richards, E. J., J. A. McGirr, J. R. Wang, M. E. St. John, J. W. Poelstra, M. J. Solano, D. C. O'Connell, et al. 2021. "A Vertebrate Adaptive Radiation Is Assembled from an Ancient and Disjunct Spatiotemporal Landscape." *Proceedings of the National Academy of Sciences* 118: e2011811118.

Selz, O. M., and O. Seehausen. 2019. "Interspecific Hybridization Can Generate Functional Novelty in Cichlid Fish." *Proceedings of the Royal Society B: Biological Sciences* 286: 20191621.

Selz, O. M., R. Thommen, M. E. Maan, and O. Seehausen. 2014. "Behavioural Isolation May Facilitate Homoploid Hybrid Speciation in Cichlid Fish." *Journal of Evolutionary Biology* 27: 275–289.

Hybridization and Introgression Could Have Accelerated Adaptive Radiation. . . . Did They?

Blaxter, M., J. M. Archibald, A. K. Childers, J. A. Coddington, K. A. Crandall, F. Di Palma, R. Durbin, et al. 2022. "Why Sequence All Eukaryotes?" *Proceedings of the National Academy of Sciences* 119: p.e2115636118.

Ivory, S. J., M. W. Blome, J. W. King, M. M. McGlue, J. E. Cole, and A. S. Cohen. 2016. "Environmental Change Explains Cichlid Adaptive Radiation at Lake Malawi over the Past 1.2 Million Years." *Proceedings of the National Academy of Sciences* 113: 11895–11900.

Malinsky, M., H. Svardal, A. M. Tyers, E. A. Miska, M. J. Genner, G. F. Turner, and R. Durbin. 2018. "Whole-Genome Sequences of Malawi Cichlids Reveal Multiple Radiations Interconnected by Gene Flow." *Nature Ecology and Evolution* 2: 1940–1955.

McGee, M. D., S. R. Borstein, J. I. Meier, D. A. Marques, S. Mwaiko, A. Taabu, M. A. Kishe, et al. 2020. "The Ecological and Genomic Basis of Explosive Adaptive Radiation." *Nature* 586: 75–79.

Meier, J. I., D. A. Marques, S. Mwaiko, C. E. Wagner, L. Excoffier, and O. Seehausen. 2017. "Ancient Hybridization Fuels Rapid Cichlid Fish Adaptive Radiations." *Nature Communications* 8: 1–11.

Meyer, A., T. D. Kocher, P. Basasibwaki, and A. C. Wilson. 1990. "Monophyletic Origin of Lake Victoria Cichlid Fishes Suggested by Mitochondrial DNA Sequences." *Nature* 347: 550–553.

Ronco, F., M. Matschiner, A. Böhne, A. Boila, H. H. Büscher, A. El Taher, A. Indermaur, et al. 2020. "Drivers and Dynamics of a Massive Adaptive Radiation in Cichlid Fishes." *Nature* 589: 76–81.

Seehausen, O., E. Koetsier, M. V. Schneider, L. J. Chapman, C. A. Chapman, M. E. Knight, G. F. Turner, et al. 2003. "Nuclear Markers Reveal Unexpected Genetic Variation and a Congolese-Nilotic Origin of the Lake Victoria Cichlid Species Flock." *Proceedings of the Royal Society B: Biological Sciences* 270: 129–137.

Svardal, H., F. X. Quah, M. Malinsky, B. P. Ngatunga, E. A. Miska, W. Salzburger, M. J. Genner, et al. 2020. "Ancestral Hybridization Facilitated Species Diversification in the Lake Malawi Cichlid Fish Adaptive Radiation." *Molecular Biology and Evolution* 37: 1100–1113.

Svardal, H., W. Salzburger, and M. Malinsky. 2021. "Genetic Variation and Hybridization in Evolutionary Radiations of Cichlid Fishes." *Annual Review of Animal Biosciences* 9: 55–79.

Takahashi, T., and E. Moreno. 2015. "A RAD-Based Phylogenetics for *Orestias* Fishes from Lake Titicaca." *Molecular Phylogenetics and Evolution* 93: 307–317.

The Transporter Hypothesis

Colosimo, P. F., K. E. Hosemann, S. Balabhadra, G. Villarreal, M. Dickson, J. Grimwood, J. Schmutz, et al. 2005. "Widespread Parallel Evolution in Sticklebacks by Repeated Fixation of Ectodysplasin Alleles." *Science* 307: 1928–1933.

McKinnon, J. S., J. Kitano, and N. Aubin-Horth. 2019. "*Gasterosteus, Anolis, Mus*, and More: The Changing Roles of Vertebrate Models in Evolution and Behaviour." *Evolutionary Ecology Research* 20: 1–25.

Nelson, T. C., and W. A. Cresko. 2018. "Ancient Genomic Variation Underlies Repeated Ecological Adaptation in Young Stickleback Populations." *Evolution Letters* 2: 9–21.

Roberts Kingman, G. A., D. N. Vyas, F. C. Jones, S. D. Brady, H. I. Chen, K. Reid, M. Milhaven, et al. 2021. "Predicting Future from Past: The Genomic Basis of Recurrent and Rapid Stickleback Evolution." *Science Advances* 7: eabg5285.

Schluter, D., and G. L. Conte. 2009. "Genetics and Ecological Speciation." *Proceedings of the National Academy of Sciences* 106: 9955–9962.

CHAPTER 7

Stager, J. C., and T. C. Johnson. 2008. "The Late Pleistocene Desiccation of Lake Victoria and the Origin of Its Endemic Biota." *Hydrobiologia* 596: 5–16.

Sediments Introduction

Cohen, A. S. 2012. "Scientific Drilling and Biological Evolution in Ancient Lakes: Lessons Learned and Recommendations for the Future." *Hydrobiologia* 682: 3–25.

Cuenca-Cambronero, M., C. J. Courtney-Mustaphi, R. Greenway, O. Heiri, C. M. Hudson, L. King, K. D. Lemmen, et al. 2022. "An Integrative Paleolimnological Approach for Studying Evolutionary Processes." *Trends in Ecology & Evolution* 37: 488–496.

Parducci, L., K. D. Bennett, G. F. Ficetola, I. G. Alsos, Y. Suyama, J. R. Wood, and M. Winther Pedersen. 2017. "Ancient Plant DNA in Lake Sediments." *New Phytologist* 214: 924–942.

Russell, J. M., P. Barker, A. Cohen, S. Ivory, I. Kimirei, C. Lane, M. Leng, et al. 2020. "ICDP Workshop on the Lake Tanganyika Scientific Drilling Project: A Late Miocene–Present Record of Climate, Rifting, and Ecosystem Evolution from the World's Oldest Tropical Lake." *Scientific Drilling* 27: 53–60.

Wang, Y., M. W. Pedersen, I. G. Alsos, B. De Sanctis, F. Racimo, A. Prohaska, E. Coissac, et al. 2021. "Late Quaternary Dynamics of Arctic Biota from Ancient Environmental Genomics." *Nature* 600: 86–92.

Wilke, T., B. Wagner, B. Van Bocxlaer, C. Albrecht, D. Ariztegui, D. Delicado, A. Francke, et al. 2016. "Scientific Drilling Projects in Ancient Lakes: Integrating Geological and Biological Histories." *Global and Planetary Change* 143: 118–151.

Almanacs in the Mud

Ageli, M. K., P. B. Hamilton, A. J. Bramburger, R. P. Weidman, Z. Song, J. Russell, H. Vogel, et al. 2022. "Benthic-Pelagic State Changes in the Primary Trophic Level of an Ancient Tropical Lake." *Palaeogeography, Palaeoclimatology, Palaeoecology* 594: 110937.

Crowe, S. A., A. H. O'Neill, S. Katsev, P. Hehanussa, G. D. Haffner, B. Sundby, A. Mucci, and D. A. Fowle. 2008. "The Biogeochemistry of Tropical Lakes: A Case Study from Lake Matano, Indonesia." *Limnology and Oceanography* 53: 319–331.

Doble, C. J., H. Hipperson, W. Salzburger, G. J. Horsburgh, C. Mwita, D. J. Murrell, and J. J. Day. 2020. "Testing the Performance of Environmental DNA Metabarcoding for Surveying Highly Diverse Tropical Fish Communities: A Case Study from Lake Tanganyika." *Environmental DNA* 2: 24–41.

Ivory, S. J., M. W. Blome, J. W. King, M. M. McGlue, J. E. Cole, and A. S. Cohen. 2016. "Environmental Change Explains Cichlid Adaptive Radiation at Lake Malawi over the Past 1.2 Million Years." *Proceedings of the National Academy of Sciences* 113: 11895–11900.

Johnson, T. C., C. A. Scholz, M. R. Talbot, K. Kelts, R. D. Ricketts, G. Ngobi, K. Beuning, et al. 1996. "Late Pleistocene Desiccation of Lake Victoria and Rapid Evolution of Cichlid Fishes." *Science* 273: 1091–1093.

Malinsky, M., and W. Salzburger. 2016. "Environmental Context for Understanding the Iconic Adaptive Radiation of Cichlid Fishes in Lake Malawi." *Proceedings of the National Academy of Sciences* 113: 11654–11656.

Russell, J. M., S. Bijaksana, H. Vogel, M. Melles, J. Kallmeyer, D. Ariztegui, S. Crowe, et al. 2016. "The Towuti Drilling Project: Paleoenvironments, Biological Evolution, and Geomicrobiology of a Tropical Pacific Lake." *Scientific Drilling* 21: 29–40.

Russell, J. M., H. Vogel, S. Bijaksana, M. Melles, A. Deino, A. Hafidz, D. Haffner, et al. 2020. "The Late Quaternary Tectonic, Biogeochemical, and Environmental Evolution of Ferruginous Lake Towuti, Indonesia." *Palaeogeography, Palaeoclimatology, Palaeoecology* 556: 109905.

Stelbrink, B., I. Stöger, R. K. Hadiaty, U. K. Schliewen, and F. Herder. 2014. "Age Estimates for an Adaptive Lake Fish Radiation, Its Mitochondrial Introgression, and an Unexpected Sister Group: Sailfin Silversides of the Malili Lakes System in Sulawesi." *BMC Evolutionary Biology* 14: 1–14.

Bibles in the Mud

Cohen, A. S. 2018. "The Past Is a Key to the Future: Lessons Paleoecological Data Can Provide for Management of the African Great Lakes." *Journal of Great Lakes Research* 44: 1142–1153.

Conith, A. J., and R. C. Albertson. 2021. "The Cichlid Oral and Pharyngeal Jaws Are Evolutionarily and Genetically Coupled." *Nature Communications* 12: 1–11.

Kuwae, M., H. Tamai, M. K. Sakata, T. Minamoto, and Y. Suzuki. 2020. "Sedimentary DNA Tracks Decadal-Centennial Changes in Fish Abundance." *Communications Biology* 3: 1–12.

Liem, K. F. 1973. "Evolutionary Strategies and Morphological Innovations: Cichlid Pharyngeal Jaws." *Systematic Zoology* 22: 425–441.

Muschick, M., E. Jemmi, N. Lengacher, S. Hänsch, N. Wales, M. Kishe, S. Mwaiko, et al. 2023. "Ancient DNA Is Preserved in Fish Fossils from Tropical Lake Sediments." Preprint.

Muschick, M., J. M. Russell, E. Jemmi, J. Walker, K. M. Stewart, A. M. Murray, N. Dubois, et al. 2018. "Arrival Order and Release from Competition Does Not Explain Why Haplochromine Cichlids Radiated in Lake Victoria." *Proceedings of the Royal Society B: Biological Sciences* 285: 20180462.

Olajos, F., F. Bokma, P. Bartels, E. Myrstener, J. Rydberg, G. Öhlund, R. Bindler, et al. 2018. "Estimating Species Colonization Dates Using DNA in Lake Sediment." *Methods in Ecology and Evolution* 9: 535–543.

Ronco, F., and W. Salzburger. 2021. "Tracing Evolutionary Decoupling of Oral and Pharyngeal Jaws in Cichlid Fishes." *Evolution Letters* 5: 625–635.

Stelbrink, B., E. Jovanovska, Z. Levkov, N. Ognjanova-Rumenova, T. Wilke, and C. Albrecht. 2018. "Diatoms Do Radiate: Evidence for a Freshwater Species Flock." *Journal of Evolutionary Biology* 31: 1969–1975.

Wilke, T., T. Hauffe, E. Jovanovska, A. Cvetkoska, T. Donders, K. Ekschmitt, A. Francke, et al. 2020. "Deep Drilling Reveals Massive Shifts in Evolutionary Dynamics after Formation of Ancient Ecosystem." *Science Advances* 6: eabb2943.

CHAPTER 8

Introducing the Blue Eye

Emlen, D., and C. Zimmer. 2020. *Making Sense of Life*. 3rd ed. New York: Macmillan.

Godefroit, P., Y. L. Bolotsky, and J. Van Itterbeeck. 2004. "The Lambeosaurine Dinosaur *Amurosaurus riabinini*, from the Maastrichtian of Far Eastern Russia." *Acta Palaeontologica Polonica* 49: 585–618.

Ksepka, D. T. 2014. "Flight Performance of the Largest Volant Bird." *Proceedings of the National Academy of Sciences* 111: 10624–10629.

Mats, V. D., D. Y. Shcherbakov, and I. M. Efimova. 2011. "Late Cretaceous-Cenozoic History of the Lake Baikal Depression and Formation of Its Unique Biodiversity." *Stratigraphy and Geological Correlation* 19: 404–423.

Mitchell, K. J., A. Scanferla, E. Soibelzon, R. Bonini, J. Ochoa, and A. Cooper. 2016. "Ancient DNA from the Extinct South American Giant Glyptodont *Doedicurus* sp. (Xenarthra: Glyptodontidae) Reveals That Glyptodonts Evolved from Eocene Armadillos." *Molecular Ecology* 25: 3499–3508.

O'Brien, C. L., M. Huber, E. Thomas, M. Pagani, J. R. Super, L. E. Elder, and P. M. Hull. 2020. "The Enigma of Oligocene Climate and Global Surface Temperature Evolution." *Proceedings of the National Academy of Sciences* 117: 25302–25309.

Reis, M. D., G. F. Gunnell, J. Barba-Montoya, A. Wilkins, Z. Yang, and A. D. Yoder. 2018. "Using Phylogenomic Data to Explore the Effects of Relaxed Clocks and Calibration Strategies on Divergence Time Estimation: Primates as a Test Case." *Systematic Biology* 67: 594–615.

Sherbakov, D. Y. 1999. "Molecular Phylogenetic Studies on the Origin of Biodiversity in Lake Baikal." *Trends in Ecology & Evolution* 14: 92–95.

Offshore Baikal

Bowman, L. L., Jr., D. J. MacGuigan, M. E. Gorchels, M. M. Cahillane, and M. V. Moore. 2019. "Revealing Paraphyly and Placement of Extinct Species within *Epischura* (Copepoda: Calanoida) Using Molecular Data and Quantitative Morphometrics." *Molecular Phylogenetics and Evolution* 140: 106578.

Dean, C. 2008. "Family Science Project Yields Surprising Data about a Siberian Lake," *New York Times*. May 6. https://www.nytimes.com/2008/05/06/science/earth/06lake.html.

Hampton, S. E., R. Lyubov, M. V. Izmest'eva, S. L. Moore, B. D. Katz, and E. A. Silow. 2008. "Sixty Years of Environmental Change in the World's Largest Freshwater Lake—Lake Baikal, Siberia." *Global Change Biology* 14: 1947–1958.

Kozhov, M. 1963. *Lake Baikal and Its Life*. The Hague: W. Junk Publishers.

Kozhova, O. M., and L. R. Izmest'eva. 1998. *Lake Baikal Evolution and Biodiversity*. Leiden: Backhuys Publishers.

Ozersky, T., T. Nakov, S. E. Hampton, N. L. Rodenhouse, K. H. Woo, K. Shchapov, K. Wright, et al. 2020. "Hot and Sick? Impacts of Warming and a Parasite on the Dominant Zooplankter of Lake Baikal." *Limnology and Oceanography* 65: 2772–2786.

Offshore Baikal, Gammarid Amphipods

Burskaia, V., S. Naumenko, M. Schelkunov, D. Bedulina, T. Neretina, A. Kondrashov, L. Yampolsky, and G. A. Bazykin. 2020. "Excessive Parallelism in Protein Evolution of Lake Baikal Amphipod Species Flock." *Genome Biology and Evolution* 12: 1493–1503.

Copilaş-Ciocianu, D., and D. Sidorov. 2022. "Taxonomic, Ecological and Morphological Diversity of Ponto-Caspian Gammaridean Amphipods: A Review." *Organisms Diversity & Evolution* 22: 285–315.

Hou, Z., and B. Sket. 2016. "A Review of Gammaridae (Crustacea: Amphipoda): The Family Extent, Its Evolutionary History, and Taxonomic Redefinition of Genera." *Zoological Journal of the Linnean Society* 176: 323–348.

Hou, Z., B. Sket, C. Fišer, and S. Li. 2011. "Eocene Habitat Shift from Saline to Freshwater Promoted Tethyan Amphipod Diversification." *Proceedings of the National Academy of Sciences* 108: 14533–14538.

Macdonald, K. S, III, L. Yampolsky, and J. E. Duffy. 2005. "Molecular and Morphological Evolution of the Amphipod Radiation of Lake Baikal." *Molecular Phylogenetics and Evolution* 35: 323–343.

Melnik, N. G., O. A. Timoshkin, V. G. Sideleva, S. V. Pushkin, and V. S. Mamylov. 1993. "Hydroacoustic Measurement of the Density of Baikal Macrozooplankter *Macrohectopus branickii*." *Limnology and Oceanography* 38: 425–434.

Moore, M. V., B. T. De Stasio Jr., K. N. Huizenga, and E. A. Silow. 2019. "Trophic Coupling of the Microbial and the Classical Food Web in Lake Baikal, Siberia." *Freshwater Biology* 64: 138–151.

Moskalenko, V. N., T. V. Neretina, and L. Y. Yampolsky. 2020. "To the Origin of Lake Baikal Endemic Gammarid Radiations, with Description of Two New *Eulimnogammarus* spp." *Zootaxa* 4766: 457–471.

Naumenko, S. A., M. D. Logacheva, N. V. Popova, A. V. Klepikova, A. A. Penin, G. A. Bazykin, A. E. Etingova, et al. 2017. "Transcriptome-Based Phylogeny of Endemic Lake Baikal Amphipod Species Flock: Fast Speciation Accompanied by Frequent Episodes of Positive Selection." *Molecular Ecology* 26: 536–553.

Naumova, E. Y., I. Y. Zaidykov, and M. M. Makarov. 2020. "Recent Quantitative Values of *Macrohectopus branickii* (Dyb.) (Amphipoda) from Lake Baikal." *Journal of Great Lakes Research* 46: 48–52.

Rudstam, L. G., N. G. Melnik, O. A. Timoshkin, S. Hansson, S. V. Pushkin, and V. Nemov. 1992. "Daily Dynamics of an Aggregation of *Macrohectopus branickii* (Dyb.) (Amphipoda, Gammaridae) in Barguzin Bay, Lake Baikal, Russia." *Journal of Great Lakes Research* 18: 286–297.

Takhteev, V. V. 2000. "Trends in the Evolution of Baikal Amphipods and Evolutionary Parallels with Some Marine Malacostracan Faunas." *Advances in Ecological Research* 31: 197–220.

Offshore Baikal, Sculpins

Goto, A., R. Yokoyama, and V. G. Sideleva. 2015. "Evolutionary Diversification in Freshwater Sculpins (Cottoidea): A Review of Two Major Adaptive Radiations." *Environmental Biology of Fishes* 98: 307–335.

Ikusemiju, K. 1975. "Aspects of the Ecology and Life History of the Sculpin, *Coitus aleuticus* (Gilbert), in Lake Washington." *Journal of Fish Biology* 7: 235–245.

Kontula, T., S. V. Kirilchik, and R. Väinölä. 2003. "Endemic Diversification of the Monophyletic Cottoid Fish Species Flock in Lake Baikal Explored with mtDNA Sequencing." *Molecular Phylogenetics and Evolution* 27: 143–155.

Miyasaka, H., Y. V. Dzyuba, M. Genkai-Kato, S. Ito, A. Kohzu, P. N. Anoshko, I. V. Khanayev, et al. 2006. "Feeding Ecology of Two Planktonic Sculpins, *Comephorus baicalensis* and *Comephorus dybowskii* (Comephoridae), in Lake Baikal." *Ichthyological Research* 53: 419–422.

Sideleva, V. G. 1996. "Comparative Character of the Deep-Water and Inshore Cottoid Fishes Endemic to Lake Baikal." *Journal of Fish Biology* 49: 192–206.

Sideleva, V. G. 2000. "The Ichthyofauna of Lake Baikal, with Special Reference to Its Zoogeographical Relations." *Advances in Ecological Research* 31: 81–96.

St. John, C. A., T. J. Buser, V. E. Kee, S. Kirilchik, B. Bogdanov, D. Neely, M. Sandel, and A. Aguilar. 2021. "Diversification along a Benthic to Pelagic Gradient Contributes to Fish Diversity in the World's Largest Lake (Lake Baikal, Russia)." *Molecular Ecology* 31: 238-251.

Woodruff, P. E., and E. B. Taylor. 2013. "Assessing the Distinctiveness of the Cultus Pygmy Sculpin, a Threatened Endemic, from the Widespread Coastrange Sculpin *Cottus aleuticus*." *Endangered Species Research* 20: 181–194.

Yoshii, K., N. G. Melnik, O. A. Timoshkin, N. A. Bondarenko, P. N. Anoshko, T. Yoshioka, and E. Wada. 1999. "Stable Isotope Analyses of the Pelagic Food Web in Lake Baikal." *Limnology and Oceanography* 44: 502–511.

Offshore Baikal, Seals

Fulton, T. L., and C. Strobeck. 2010. "Multiple Fossil Calibrations, Nuclear Loci and Mitochondrial Genomes Provide New Insight into Biogeography and Divergence Timing for True Seals (Phocidae, Pinnipedia)." *Journal of Biogeography* 37: 814–829.

Klages, N. T. W., and V. G. Cockcroft. 1990. "Feeding Behaviour of a Captive Crabeater Seal." *Polar Biology* 10: 403–404.

Kozhova, O. M., and L. R. Izmest'eva. 1998. *Lake Baikal Evolution and Biodiversity*. Leiden: Backhuys Publishers.

Palo, J. U., and R. Väinölä. 2006. "The Enigma of the Landlocked Baikal and Caspian Seals Addressed through Phylogeny of Phocine Mitochondrial Sequences." *Biological Journal of the Linnean Society* 88: 61–72.

Watanabe, Y. Y., E. A. Baranov, and N. Miyazaki. 2020. "Ultrahigh Foraging Rates of Baikal Seals Make Tiny Endemic Amphipods Profitable in Lake Baikal." *Proceedings of the National Academy of Sciences* 117: 31242–31248.

The Abyss: Desert and Oasis

Kozhova, O. M., and L. R. Izmest'eva. 1998. *Lake Baikal Evolution and Biodiversity*. Leiden: Backhuys Publishers.

Papoucheva, E., V. Proviz, C. Lambkin, B. Goddeeris, and A. Blinov. 2003. "Phylogeny of the Endemic Baikalian *Sergentia* (Chironomidae, Diptera)." *Molecular Phylogenetics and Evolution* 29: 120–125.

Sideleva, V. G. 1996. "Comparative Character of the Deep-Water and Inshore Cottoid Fishes Endemic to Lake Baikal." *Journal of Fish Biology* 49: 192–206.

Sideleva, V. G. 2000. "The Ichthyofauna of Lake Baikal, with Special Reference to Its Zoogeographical Relations." *Advances in Ecological Research* 31: 81–96.

Sideleva, V. G. 2016. "Communities of the Cottoid Fish (Cottoidei) in the Areas of Hydrothermal Vents and Cold Seeps of the Abyssal Zone of Baikal Lake." *Journal of Ichthyology* 56: 694–701.

Sideleva, V. G., and V. A. Fialkov. 2014. "Fauna of Cottoid Fish (Cottoidei) in the Area of Methane Seep in the Abyssal of Lake Baikal." *Doklady Biological Sciences* 459: 351–353.

Sitnikova, T. Y., I. V. Mekhanikova, V. G. Sideleva, S. I. Kiyashko, T. V. Naumova, T. I. Zemskaya, and O. M. Khlystov. 2017. "Trophic Relationships between Macroinvertebrates and Fish in St. Petersburg Methane Seep Community in Abyssal Zone of Lake Baikal." *Contemporary Problems of Ecology* 10: 147–156.

Tiercelin, J. J., C. Pflumio, M. Castrec, J. Boulégue, P. Gente, J. Rolet, C. Coussement, et al. 1993. "Hydrothermal Vents in Lake Tanganyika, East African Rift System." *Geology* 21: 499–502.

Zemskaya, T. I., T. Y. Sitnikova, and O. M. Khlystov. 2014. "Baikal Deep-zone Studies." *Herald of the Russian Academy of Sciences* 84: 183–187.

Zemskaya, T. I., T. Y. Sitnikova, S. I. Kiyashko, G. V. Kalmychkov, T. V. Pogodaeva, I. V. Mekhanikova, T. V. Naumova, et al. 2012. "Faunal Communities at Sites of Gas-and Oil-Bearing Fluids in Lake Baikal." *Geo-Marine Letters* 32: 437–451.

Baikal in the Anthropocene

Hampton, S. E., R. Lyubov, M. V. Izmest'eva, S. L. Moore, B. D. Katz, and E. A. Silow. 2008. "Sixty Years of Environmental Change in the World's Largest Freshwater Lake–Lake Baikal, Siberia." *Global Change Biology* 14: 1947–1958.

Swann, G. E. A., V. N. Panizzo, S. Piccolroaz, V. Pashley, M. S. A. Horstwood, S. Roberts, E. Vologina, et al. 2020. "Changing Nutrient Cycling in Lake Baikal, the World's Oldest Lake." *Proceedings of the National Academy of Sciences* 117: 27211–27217.

Wollrab, S., L. Izmest'eva, S. E. Hampton, E. A. Silow, E. Litchman, and C. A. Klausmeier. 2021. "Climate Change–Driven Regime Shifts in a Planktonic Food Web." *American Naturalist* 197: 281–295.

CHAPTER 9

Introduction, Islands

Fernández-Palacios, J. M., H. Kreft, S. D. H. Irl, S. Norder, C. Ah-Peng, P. A. V. Borges, K. C. Burns, et al. 2021. "Scientists' Warning: The Outstanding Biodiversity of Islands Is in Peril." *Global Ecology and Conservation* 31: e01847.

Nogué, S., A. M. C. Santos, H. J. B. Birks, S. Björck, A. Castilla-Beltrán, S. Connor, E. J. de Boer, et al. 2021. "The Human Dimension of Biodiversity Changes on Islands." *Science* 372: 488–491.

Tickner, D., J. J. Opperman, R. Abell, M. Acreman, A. H. Arthington, S. E. Bunn, S. J. Cooke, et al. 2020. "Bending the Curve of Global Freshwater Biodiversity Loss: An Emergency Recovery Plan." *BioScience* 70: 330–342.

Introduction, Lanao

Abdulmalik-Labe, O. P., and J. P. Quilang. 2019. "DNA Barcoding of Fishes from Lake Lanao, Philippines." *Mitochondrial DNA Part B* 4: 1890–1894.

Herre, A. W. C. T. 1933. "The Fishes of Lake Lanao: A Problem in Evolution." *American Naturalist* 67: 154–162.

Ismail, G. B., D. B. Sampson, and D. L. G. Noakes. 2014. "The Status of Lake Lanao Endemic Cyprinids (*Puntius* Species) and Their Conservation." *Environmental Biology of Fishes* 97: 425–434.

Myers, G. S. 1960. "The Endemic Fish Fauna of Lake Lanao, and the Evolution of Higher Taxonomic Categories." *Evolution* 14: 323–333.

Stelbrink, B., T. von Rintelen, C. Albrecht, C. Clewing, and P. O. Naga. 2019. "Forgotten for Decades: Lake Lanao and the Genetic Assessment of Its Mollusc Diversity." *Hydrobiologia* 843: 31–49.

Value

Coutinho, F. H., P. J. Cabello-Yeves, R. Gonzalez-Serrano, R. Rosselli, M. López-Pérez, T. I. Zemskaya, A. S. Zakharenko, et al. 2020. "New Viral Biogeochemical Roles Revealed through Metagenomic Analysis of Lake Baikal." *Microbiome* 8: 1–15.

European Environmental Agency. n.d. Common International Classification of Ecosystem Services Version 5.1. https://cices.eu.

Howes, M. J. R., C. L. Quave, J. Collemare, E. C. Tatsis, D. Twilley, E. Lulekal, A. Farlow, et al. 2020. "Molecules from Nature: Reconciling Biodiversity Conservation and Global Healthcare Imperatives for Sustainable Use of Medicinal Plants and Fungi." *Plants, People, Planet* 2: 463–481.

Koonin, E. V., and K. S. Makarova. 2019. "Origins and Evolution of CRISPR-Cas Systems." *Philosophical Transactions of the Royal Society B* 374: 20180087.

Sterner, R. W., B. Keeler, S. Polasky, R. Poudel, K. Rhude, and M. Rogers. 2020. "Ecosystem Services of Earth's Largest Freshwater Lakes." *Ecosystem Services* 41: 101046.

Extinction, Ecology, and HIPPO

Aladin, N. V., V. I. Gontar, L. V. Zhakova, I. S. Plotnikov, A. O. Smurov, P. Rzymski, and P. Klimaszyk. 2019. "The Zoocenosis of the Aral Sea: Six Decades of Fast-Paced Change." *Environmental Science and Pollution Research* 26: 2228–2237.

Cohen, A. S., E. L. Gergurich, B. M. Kraemer, M. M. McGlue, P. B. McIntyre, J. M. Russell, J. D. Simmons, and P. W. Swarzenski. 2016. "Climate Warming Reduces Fish Production and Benthic Habitat in Lake Tanganyika, One of the Most Biodiverse Freshwater Ecosystems." *Proceedings of the National Academy of Sciences* 113: 9563–9568.

Hampton, S. E., S. McGowan, T. Ozersky, S. G. P. Virdis, T. T. Vu, T. L. Spanbauer, B. M. Kraemer, et al. 2018. "Recent Ecological Change in Ancient Lakes." *Limnology and Oceanography* 63: 2277–2304.

Kamulali, T. M., M. M. McGlue, J. R. Stone, I. A. Kimirei, P. J. Goodman, and A. S. Cohen. 2022. "Paleoecological Analysis of Holocene Sediment Cores from the Southern Basin of Lake Tanganyika: Implications for the Future of the Fishery in One of Africa's Largest Lakes." *Journal of Paleolimnology* 67: 17–34.

Micklin, P. 2007. "The Aral Sea Disaster." *Annual Review of Earth and Planetary Sciences* 35: 47–72.

Plotnikov, I. S., Z. K. Ermakhanov, N. V. Aladin, and P. Micklin. 2016. "Modern State of the Small (Northern) Aral Sea Fauna." *Lakes and Reservoirs: Research and Management* 21: 315–328.

Verheyen, E., R. Abila, P. Akoll, C. Albertson, D. Antunes, T. Banda, R. Bills, et al. 2016. "Oil Extraction Imperils Africa's Great Lakes." *Science* 354: 561–562.

Species Introductions, Harvesting, and Interactions with Other Impacts

Dönz, C. J., and O. Seehausen. 2020. "Rediscovery of a Presumed Extinct Species, *Salvelinus profundus*, after Re-Oligotrophication." *Ecology* 101: e03065.

Goldschmidt, T. 1996. *Darwin's Dreampond.* Cambridge, MA: MIT Press.

Haase, M., T. von Rintelen, B. Harting, R. Marwoto, and M. Glaubrecht. 2023. "New Species from a 'Lost World': Sulawesidrobia (Caenogastropoda, Tateidae) from Ancient Lake Matano, Sulawesi, Indonesia." *European Journal of Taxonomy* 864: 77–103.

Herder, F., J. Möhring, J. M. Flury, I. V. Utama, L. Wantania, D. Wowor, F. B. Boneka, et al. 2022. "More Non-Native Fish Species than Natives, and an Invasion of Malawi Cichlids, in Ancient Lake Poso, Sulawesi, Indonesia." *Aquatic Invasions* 17: 72–91.

Herder, F., U. K. Schliewen, M. F. Geiger, R. K. Hadiaty, S. M. Gray, J. S. McKinnon, R. P. Walter, and J. Pfaender. 2012. "Alien Invasion in Wallace's Dreamponds: Records of the Hybridogenic 'Flowerhorn' Cichlid in Lake Matano, with an Annotated Checklist of Fish Species Introduced to the Malili Lakes System in Sulawesi." *Aquatic Invasions* 7: 521–535.

Hilgers, L., F. Herder, R. K. Hadiaty, and J. Pfaender. 2018. "Alien Attack: Trophic Interactions of Flowerhorn Cichlids with Endemics of Ancient Lake Matano (Sulawesi, Indonesia)." *Evolutionary Ecology Research* 19: 561–574.

McGee, M. D., S. R. Borstein, R. Y. Neches, H. H. Buescher, O. Seehausen, and P. C. Wainwright. 2015. "A Pharyngeal Jaw Evolutionary Innovation Facilitated Extinction in Lake Victoria Cichlids." *Science* 350: 1077–1079.

Nasution, S. H., A. M. Muchlis, and H. T. Cinnawara. 2022. "The Abundance of Alien Fish Species Flowerhorn (*Cichlasoma trimaculatum*) (GÜNTHER, 1867) in Its Fishing Ground Area at Lake Mahalona, South Sulawesi." *IOP Conference Series: Earth and Environmental Science* 1036: 012103.

Natugonza, V., L. Musinguzi, M. A. Kishe, J. C. van Rijssel, O. Seehausen, and R. Ogutu-Ohwayo. 2021. "The Consequences of Anthropogenic Stressors on Cichlid Fish Communities: Revisiting Lakes Victoria, Kyoga, and Nabugabo." In *The Behavior, Ecology and Evolution of Cichlid Fishes*, edited by M. E. Abate and D. L. G. Noakes, 217–246. Dordrecht: Springer.

Sentosa, A. A., and D. A. Hedianto. 2020. "Gillnets Selectivity and Effectivity for Controlling Invasive Fish Species in Lake Matano, South Sulawesi." *IOP Conference Series: Earth and Environmental Science* 535: 012039.

Prospects

Albert, J. S., G. Destouni, S. M. Duke-Sylvester, A. E. Magurran, T. Oberdorff, R. E. Reis, K. O. Winemiller, and W. J. Ripple. 2021. "Scientists' Warning to Humanity on the Freshwater Biodiversity Crisis." *Ambio* 50: 85–94.

Di Franco, A., P. Thiriet, G. Di Carlo, C. Dimitriadis, P. Francour, N. L. Gutiérrez, A. J. de Grissac, et al. 2016. "Five Key Attributes Can Increase Marine Protected Areas Performance for Small-Scale Fisheries Management." *Scientific Reports* 6: 1–9.

Jenny, J. P., O. Anneville, F. Arnaud, Y. Baulaz, D. Bouffard, I. Domaizon, S. A. Bocaniov, et al. 2020. "Scientists' Warning to Humanity: Rapid Degradation of the World's Large Lakes." *Journal of Great Lakes Research* 46: 686–702.

Ripple, W. J., C. Wolf, T. M. Newsome, J. W. Gregg, T. M. Lenton, I. Palomo, J. A. J. Eikelboom, et al. 2021. "World Scientists' Warning of a Climate Emergency." *BioScience* 71: 894–898.

GLOSSARY

Emlen, D., and C. Zimmer. 2020. *Making Sense of Life*. 3rd ed. New York: Macmillan.

Gray, S. M., and J. S. McKinnon. 2007. "Linking Color Polymorphism Maintenance and Speciation." *Trends in Ecology & Evolution* 22: 71–79.

Index